Space is nowhere completely en[...] vast clouds of gas and dust. Som[...] colors; others are so faint they ca[...] telescopes.  [...]t [...] by sensitive radio and infrared

Interstellar matter is the stuff out of which the Solar System, the Earth and even our own bodies was made: there is enough of it left in the Milky Way Galaxy to make ten billion more stars like the Sun. To some astronomers interstellar matter is a nuisance, since it hides and distorts their view of more distant objects in the Universe. Others view it as a cosmic laboratory whose vastness allows atoms to behave in ways that cannot be duplicated on Earth.

The fullness of space is a comprehensive account of what astronomers have learned about interstellar matter – where it comes from, what it is made of, and how it collects together to form new stars and planets. It is beautifully illustrated with photographs and computer-generated images of nebulae, dust clouds and galaxies. The text is non-technical. No prior knowledge of astronomy or physics is needed to enjoy this introduction.

The Fullness of Space

The Fullness of Space –
nebulae, stardust, and the interstellar medium

Gareth Wynn-Williams
Institute for Astronomy, University of Hawaii

Published by the Press Syndicate of the University of Cambridge
The Pitt Building, Trumpington Street, Cambridge CB2 1RP
40 West 20th Street, New York, NY 10011-4211, USA
10 Stamford Road, Oakleigh, Victoria 3166, Australia

© Cambridge University Press 1992

First published 1992

Printed and bound in Great Britain by
Butler & Tanner Ltd, Frome and London

A catalogue record of this book is available from the British Library

Library of Congress cataloging in publication data
Wynn-Williams, C. G. (C. Gareth), 1944–
  The fullness of space/Gareth Wynn-Williams.
    p.   cm.
   ISBN 0 521 35591 5   0 521 42638 3 (paperback)
   1. Astronomy. 2. Interstellar matter. 3. Cosmic dust.
  I. Title.
  QB43.2.W96   1992
  523.1′125–dc20   91-27956 CIP

ISBN 0 521 35591 5 hardback
ISBN 0 521 42638 3 paperback

For Chris, Oliver and Harry

Contents

Preface	*page* xiii
1 Lumps and spaces	1
1.1 Stars, planets and galaxies	1
1.2 Exploring the Galaxy	2
1.3 The discovery of interstellar matter	3
1.4 The emptiness of space	5
1.5 Interstellar matter, *pro* and *con*	8
2 Light and radiation	9
2.1 Electromagnetic waves	10
2.2 The speed of light	11
2.3 Frequency	12
2.4 Photons and energy	12
2.5 The transport of energy	13
2.6 Distortion of electromagnetic waves	13
2.7 Walls and windows	14
2.8 Tools of the trade: telescopes	16
3 Atoms and spectra	18
3.1 The structure of the atom	19
3.2 Energy levels and transitions	20
3.3 Spectral lines	22
3.4 The Doppler effect	24
3.5 Tools of the trade: spectroscopy	25
3.6 Molecules	26
3.7 Solids	27
3.8 Temperature	27
4 Atomic gas in the interstellar medium	30
4.1 The 21-cm line	30
4.2 Atomic hydrogen in galaxies	31
4.3 Measuring the mass of a galaxy	32
4.4 Tools of the tade: radio astronomy	34
4.5 Atomic hydrogen in the Milky Way	35
4.6 Interstellar clouds	39
4.7 Gas motions in the interstellar medium	41

5 Ionized gas in the interstellar medium 43
5.1 Photoionization 43
5.2 Exciting stars 45
5.3 Tools of the trade: astronomical photography 47
5.4 The light from an H^+ region 48
5.5 Forbidden lines from ions 50
5.6 Radio and infrared emission from H^+ regions 51
5.7 The warm ionized medium 52
5.8 Coronal gas 54

6 The other elements 55
6.1 Interstellar helium 55
6.2 Heavy elements in H^+ regions 56
6.3 Tools of the trade: ultraviolet astronomy 57
6.4 Abundances from visible and ultraviolet absorption lines 59
6.5 Abundances in the Solar System 61

7 Interstellar dust 64
7.1 Extinction and reddening 64
7.2 Scattered light 66
7.3 Infrared emission 69
7.4 Tools of the trade: infrared astronomy 72
7.5 Sizes and shapes of dust particles 74
7.6 The nature of interstellar grains 75
7.7 How much dust is there? 78

8 Molecular clouds 79
8.1 Molecular spectroscopy 79
8.2 Hydrogen molecules in transparent clouds 81
8.3 Vibrationally excited hydrogen molecules 83
8.4 Interstellar chemistry 84
8.5 Dense molecular clouds 88
8.6 Tools of the trade: millimeter-wave astronomy 92
8.7 Living in a molecular cloud 93
8.8 Molecular clouds in the Galaxy 94
8.9 Interstellar masers 95

9 Cosmic rays and magnetic fields 98
9.1 Fields and particles 99
9.2 Cosmic rays in the Galaxy 99
9.3 Gamma rays from the Galaxy 101
9.4 Tools of the trade: cosmic ray and gamma ray telescopes 102
9.5 Galactic radio emission 104
9.6 Sources of cosmic rays 105
9.7 The nature of interstellar magnetism 105
9.8 Measuring the galactic magnetic field 106

10 The origin of interstellar matter 111
10.1 The origin of the elements 111
10.2 Mass loss from stars 113
10.3 Planetary nebulae 115
10.4 Supernovae 117
10.5 Tools of the trade: X-ray telescopes 121
10.6 Novae and other binary systems 122
10.7 Formation and destruction of dust 123

11 Solving the interstellar jigsaw puzzle — 125
11.1 Towards a taxonomy — 125
11.2 Heating and cooling of the interstellar medium — 126
11.3 Coronal gas revisited — 128
11.4 Bubbles, tunnels, onions and sheets — 129
11.5 The local neighborhood — 132
11.6 The galactic halo — 133

12 The formation of stars — 135
12.1 Recognizing young stars — 136
12.2 Gravity and clouds — 138
12.3 Stability and collapse — 139
12.4 Magnetic pressure — 140
12.5 Protostars — 142
12.6 Tools of the trade: computers in astronomy — 144
12.7 The problem of rotation — 145
12.8 Outflows and winds — 146
12.9 High mass stars — 148
12.10 The fate of the interstellar medium — 150

13 The interplanetary medium — 152
13.1 Interplanetary dust — 152
13.2 The origin and fate of interplanetary dust — 155
13.3 The solar wind — 156
13.4 Interstellar gas within the Solar System — 158
13.5 The edge of the Solar System — 159

14 Geospace — 162
14.1 Gravitational equilibrium — 162
14.2 Dissociation and ionization — 164
14.3 The magnetosphere — 166

15 Intergalactic matter — 169
15.1 Fountains, bridges and starbursts — 169
15.2 Hot gas in galaxy clusters — 171
15.3 Quasar absorption lines — 174
15.4 Dark matter — 175

Appendices
A: Large and small numbers — 179
B: The metric system and related units — 180
C: Greek letters — 183
D: Wavelength, frequency and energy — 184
E: Selected chemical elements — 185
F: The Doppler effect — 186
G: Temperature, energy and pressure — 187
H: Thermal radiation — 188
I: Galactic rotation — 189
J: Atmospheres and gravity — 190
K: The magnitude scale — 192
L: Boltzmann's equation — 193
M: The Jeans criterion — 194

Suggestions for further reading — 195
Picture acknowledgements — 196
Index — 199

Preface

Why should we care about interstellar matter? The clouds of gas and dust that drift between the stars in the Milky Way are far too distant to affect our daily lives in any significant way. They present no material benefit nor threat to anyone. Most of them cannot be seen without a powerful telescope.

If interstellar matter did nothing more than quietly fill the gaps between the stars then it would merit only a footnote in the inventory of the Universe. Five billion years ago, however, one particular interstellar cloud grew sufficiently dense that it collapsed under the pressure of its own gravity. Gases falling to the center of the cloud grew hot and began to glow with the brilliance of the star we now call the Sun. Gases swirling around the new star collected together to form smaller cooler bodies made of rocks, metals and ice. One of these small bodies was the planet Earth.

Every atom that now makes up the Earth, its oceans, its atmosphere and its inhabitants was once part of the interstellar medium. Each atom in our bodies was once part of a simple molecule or a microscopic dust particle floating freely in the vast open spaces of the Milky Way Galaxy. Everything we can see was once interstellar matter. Notwithstanding their role in our ancestry, interstellar clouds stretch human imagination to its limits. They are of such vast size that light waves take years to cross from one side to the other, yet most of them are more rarified than the best vacuum we can produce on Earth.

The goal of this book is to explain what interstellar matter is, how we can observe it, and why it behaves the way it does. This goal cannot be achieved without some understanding of physics, but one of the beauties of interstellar matter is that its behavior is fundamentally simple. Light travels through the Galaxy in almost perfectly straight lines, atoms behave almost exactly as described in elementary physics textbooks, and liquids do not exist. Different kinds of matter, such as hydrogen atoms, dust grains and cosmic rays coexist with almost no interference with each other. As a result, the study of interstellar matter has so far avoided the complexities of sciences such as biochemistry, nuclear physics or mineralogy. Because the behavior of interstellar matter is simple, its story can be told without mathematics and without assuming that the reader has any formal scientific background. I would like to think that anyone with an interest in the Universe and an appreciation of logical thinking can follow the story laid out in this book.

I would also hope the book will be of some use to serious students of physics and astronomy as a broad introduction to the range of interstellar phenomena that are currently accessible to professional astronomers. I have included several mildly technical appendices for those with some background in physics, but the appendices can be safely ignored by readers who feel that equations are a barrier rather than a help to their understanding.

The book is designed to introduce the reader to the interstellar medium piece by piece. The simplest component – neutral atomic hydrogen – is discussed first. Other ingredients are then introduced gradually before any attempt is made to understand the origin and evolution of the interstellar medium as a whole. One consequence of this approach is that scientific breakthroughs are not necessarily presented in their correct chronological sequence – radio observations are introduced before visible wavelength observations, for example. I am aware that this approach may leave the reader with a poor sense of how our understanding of the nature of interstellar matter has grown and evolved over the past few decades, but I am happy to leave the writing of history to others.

The book can be divided into four parts.

- Chapters 1–3 contain introductory material which readers with a background in astronomy or physics may choose to skip.
- Chapters 4–9 catalog the ingredients of the interstellar medium. The gases, the solids, the particles and the fields are introduced in turn. In chapters 4 and 5 we deal with hydrogen – the simplest element – in its neutral and its ionized phases. In chapters 6 and 7 we add the heavier elements such as carbon and oxygen, first in their gaseous states and then as the constituents of solid dust particles. Chapter 8 covers interstellar molecules and the clouds they are found in, while chapter 9 deals with the elusive cosmic ray particles and with the magnetic fields that play a major role in controlling the motions of interstellar gases.
- Chapters 10–12 comprise the third section of the book, in which we mix these ingredients together and explore the origin, evolution and fate of interstellar matter, while learning a little about the ecology of the Galaxy.
- In the final three chapters of the book (13–15) we apply our understanding of the interstellar medium to other nearly-empty volumes of the Universe. We look at what fills space in the Solar System, in the Earth's upper atmosphere and between the galaxies. The coverage of these three topics is less comprehensive than the coverage of the interstellar medium.

This book is much more concerned with how astronomical observations are interpreted than in how they are made. A good scientist always has to understand the limitations of his or her data, however, so most chapters in this book include a short section called 'Tools of the Trade'. These sections contain a brief description of the hardware used and the technical problems faced in obtaining some of the data relevant to that chapter. Because interstellar matter can be studied by so many different methods, almost all major astronomical techniques get mentioned in this book.

Astronomy has as much jargon as any other science. I have tried to avoid the worst, but have used technical terms when the only alternative was to be long-winded, to invent a new word, or to use an unsuitable metaphor. Rather than compile a glossary I have used **bold type** for the first appearance of each new word or phrase and have explained its meaning at its initial appearance. The first listing of a word in the index will usually lead the reader to a definition.

I would like to express my gratitude to friends and colleagues who have given me advice as I wrote the book, or reviewed early drafts of its chapters.

They include Ann Boesgaard, Jeff Goldader, Martha Hanner, George Herbig, Esther Hu, David Jewitt, Jill Knapp, Simon Lilly, Marcia Neugebauer, Kris Sellgren, and Remo Tilanus.

Finally there are the scientists whose research over the past few decades has generated the material out of which this book was written. A few are mentioned specifically, but most make their appearance in these pages via their ideas and their discoveries, rather than by their names. I thank them all.

Gareth Wynn-Williams
Honolulu

1 Lumps and spaces

The briefest glance at the night sky reveals a remarkable fact about the Universe. It is extremely patchy. The light we see on a moonless night comes from bright specks we call stars and planets; between the stars we see blackness. The night sky is a vivid contrast to the daytime sky which, whether gray or blue, has a smoothness which does not disappear until the Sun goes down.

When we look at the daytime sky what we actually see is the Earth's atmosphere lit up by the Sun. Give or take the wind and the weather, this atmosphere surrounds us like a thin smooth blanket. At night, when there is no daylight to dazzle our vision, we can see right through the atmosphere to the rest of the Universe in all its glorious lumpiness. Most of astronomy, not to mention geology, biology, and all humanistic studies, is concerned with what happens in and on the lumps. But these lumps, which include the Earth, the Sun, the planets, and all the stars, occupy together less than one billion billion billionth (10^{-27}) of the total volume of the Universe. (Throughout this book the word **billion** means 1 000 000 000. The use of scientific notation to express large and small numbers is described in Appendix A.) Libraries are full of books about the lumps. This book is one of the few that is concerned with what goes on in the vast volumes of space which are *not* occupied by lumps.

1.1 Stars, planets and galaxies

Most of the visible matter in the Universe exists as **stars**. The Sun is a typical star; it is a white-hot ball of gas which radiates energy into space in the form of light and heat. It is held together by gravity, each part of the Sun exerting a pull on every other part. The Sun is hot because it contains a vast continuously-running thermonuclear reactor at its core. In this reactor hydrogen gas – the main material out of which the Sun is made – is steadily converted into another gas, helium. The power generated by these nuclear reactions keeps the Sun hot and, incidentally, provides the Earth with the warmth it needs to support life. The heat produced also maintains the gas pressure in the Sun; without the pressure the Sun would collapse to a tiny ball under the forces of its own gravity.

Our Sun is 100 times the diameter of the Earth and contains 300 000 times as much matter. It has existed for about five billion years and is expected to shine for five billion more. When compared with other stars our Sun is quite ordinary. There are stars that are larger and stars that are smaller, stars that are hotter and stars that are cooler, stars that are older and stars that are younger. For some periods of its life a star may do something remarkable, such as pulsate, eject streams of gas into space, explode, or even collapse to become a black hole. These events, though of great fascination to astronomers, are rare. The most likely role of an atom in the Universe is to be part of a star that is calmly converting hydrogen to helium. In a sense, therefore, one should think of stars as the most ordinary things in the Universe. It is the other objects, such as the Earth, that are remarkable.

The Sun has nine planets, dozens of moons, thousands of asteroids, and millions of comets in orbit around it. Together with the Sun itself they comprise the **Solar System**. Planets, moons, asteroids and comets are smaller than stars and do not generate their own nuclear power. They are warmed by the heat of the Sun. Being cooler than stars, they may contain solids and liquids, as well as gases. Astronomers spend much time studying planets and other Solar System objects because they are close to us and because they can provide us with clues about the origins of the Earth. However, the total mass of all the objects in orbit around the Sun adds up to less than 1% of the mass of the Sun itself. In terms of the light being emitted, the Sun's dominance is even greater; it radiates a billion times more light than Jupiter, the next largest object in the Solar System. There are probably planets around other stars, but they are very hard to detect, and we know very little about them. Planets play only a small role in this book.

As far as we know, stars exist only as members of **galaxies**. The galaxy in which the Sun resides is usually referred to as the **Galaxy** with a capital G, and is visible to us on dark nights as the **Milky Way**. This band of light crossing the sky is made up from the light of an estimated 100 billion individual stars which form a gigantic disk slowly spinning in space. The stars and their attendant planets move in vast orbits around the Galaxy, making one revolution in about 100 million years. Our Galaxy is classified as a **spiral galaxy** because many of its stars are concentrated in long streamers called **spiral arms**, which wind themselves around the galaxy. It is difficult to pick out the spiral arms in the Milky Way, but they are easily seen in some other galaxies. Spiral arms give galaxies their immediately-recognizable shapes, but not all galaxies possess them.

Galaxies themselves cluster together. The galaxies in our own neighborhood go by the uninspiring name of the '**Local Group**'. The local group has about 30 members, including the Milky Way, the Magellanic Clouds and the Andromeda galaxy. Some clusters of galaxies contain thousands of members. The Universe appears to be filled with a vast network of clusters of galaxies, but we are only just beginning to glimpse the shape of this network.

1.2 Exploring the Galaxy

The main subject of this book is the matter that fills the spaces between the stars in our own Galaxy. Astronomers refer to this material as the **interstellar medium**. In the main sections of this book we will discuss how the interstellar medium is observed, what it is made of, why it behaves the way it does, where it comes from, and how it makes new stars. Interstellar space is not the only kind of space, however, and in the last three chapters we will look at the **interplanetary medium**, which fills the spaces inside our Solar System, at the Earth's **upper atmosphere** (or '**geospace**') which is where most space travel takes place, and at the elusive **intergalactic medium** which may or may not fill the whole Universe.

How much can we learn by the direct exploration of space? The farthest distance ever travelled by a human astronaut is to the Moon, 400 000 km

Figure 1.1. The spiral galaxy NGC 2997. The light from a galaxy such as this comes from billions of individual stars, each of which is in orbit around the center of the galaxy. The Milky Way Galaxy would look something like this if we could view it from a sufficient distance.

away. Unmanned spacecraft equipped with television cameras and other scientific instruments have travelled much farther. In August 1989, NASA's Voyager spacecraft sent back pictures of Neptune, at that time the most distant planet in the Solar System, 30 astronomical units (more than 4 billion km) from Earth. Satellites and space probes have allowed us to explore the Earth's upper atmosphere and the interplanetary medium in much more detail than was possible before the space age, but they can provide us with little direct information about the interstellar medium or the intergalactic medium. The problem is simply one of distance; 4 billion km is an impressive trip on a human scale, but it amounts to only 1/10 000 of the distance to even the nearest star.

Some idea of the vastness of the Universe may be gained by considering a model in which everything has been scaled down by a factor of a billion. In this model the Earth would have the dimensions of a grape. The Moon would resemble a grapeseed 40 cm away while the Sun would be a 1.4 meter diameter sphere at a distance of 150 meters. Neptune would be more than 4 km away. On this one-billionth scale the nearest star would be at a distance of 40 000 km – more than the actual diameter of the Earth. One would have to travel five thousand times farther yet to reach the center of the Milky Way Galaxy, another 80 times farther to reach the next nearest spiral galaxy, and another several thousand times farther still to reach the limits of the known Universe. Clearly, finding out what space is filled with is a task for astronomers and their telescopes rather than for astronauts and their spacecraft.

1.3 The discovery of interstellar matter

Before the development of the telescope there were few signs that interstellar space was anything other than totally empty. As telescopes became more powerful in the eighteenth century, however, astronomers such as William Herschel discovered many diffuse patches of light and called them **nebulae**, from the Latin for 'cloud'. For a long time it was thought that these nebulae

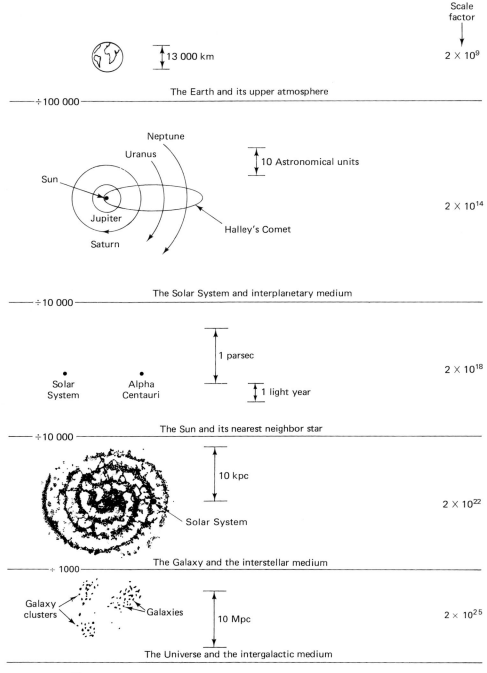

Figure 1.2. Different kinds of space, and the objects that fill them. The top panel is a sketch of the Earth and its upper atmosphere scaled down by a factor of two billion (2×10^9). Successive panels are progressively scaled down by further factors of either 10^3, 10^4 or 10^5. The scale of each picture is given in the right of each panel. An explanation of the metric system and of other units used in astronomy is given in Appendix B.

were simply clusters of faint stars, but observations of their spectra by William Huggins in the 1860s showed that some of the so-called **bright nebulae** were made up of hot gases such as hydrogen. In 1912, Vesto Slipher found other nebulae (called **reflection nebulae**) which appeared to be comprised of dust particles illuminated by a nearby star. The importance of interstellar clouds was also appreciated by Max Wolf and Edward Barnard who, in the 1920s, recognized that certain anomalously starless regions of the sky, the so-called **dark nebulae**, were clouds of dust particles which were blocking the light from the background stars.

Evidence that the whole of interstellar space, as opposed to just the nebulae, might be filled with interstellar matter was harder to obtain. In 1904, J. Hartmann found anomalies in the spectrum of the star δ Orionis and deduced that invisible patches of gas between us and the star must be interfering with the light. At about the same time it was becoming clear that there were problems in understanding the colors of many faint stars, in that distant stars often appeared to have a redder color than nearby ones. Robert Trümpler, in 1930, realized that this interstellar **reddening** must be due to a sea of dust particles that filled the Galaxy and caused distant stars to appear both redder and fainter. Trümpler's work was crucial for helping astronomers gain a proper understanding of the structure of the Milky Way Galaxy, since it then became clear that in some directions in the sky the interstellar **extinction**, or dimming of the light by a haze of dust grains, is enough to render parts of the galactic disk essentially unobservable. This problem is especially acute toward the Galaxy's center, which is hidden by an extensive array of dark clouds.

The matter that fills the spaces between the stars but which does not form part of a discrete nebula is now known as the **diffuse interstellar medium**, but it was not until the 1950s that its main constituent – atomic hydrogen gas – was first directly detected. It was only in the 1970s that the largest agglomerations of interstellar matter – the **giant molecular clouds** – were discovered. Developments in radio, infrared and ultraviolet astronomy in the past 40 years have revealed an ever richer variety of interstellar phenomena. We now know that interstellar gas and interstellar dust almost always exist together and that the different appearances of bright nebulae, dark nebulae and the diffuse interstellar medium are due to factors such as the density of the gas and its proximity to bright stars. The interplay between interstellar matter and its environment is one of the major themes of this book.

1.4 The emptiness of space

We normally think of a vacuum as a region of space from which all gas has been removed. No vacuum is perfect, however, and there are enormous differences in the degrees of emptiness of what may be loosely described as empty space. A simple way of characterizing a vacuum or a gas is by the concentration (or particle density) of the atoms in it. One cubic centimeter of air at sea level contains 50 billion billion (5×10^{19}) atoms. As one moves up in the Earth's atmosphere the density gets progressively thinner. The Space Shuttle usually flies in an orbit about 400 km above the Earth's surface. Although we refer to this region as 'space' it is actually still part of the Earth's upper atmosphere; the gas at these altitudes has a particle density of around 10^7 atoms cm^{-3} – trillions of times less than at sea level. In interstellar space, however, the densities are millions of times lower still; the average particle density in the disk of the Galaxy is around 1 atom cm^{-3}. This density is thousands of times less that that of even the best laboratory-produced vacuum on Earth; if one could pass from the best

Figure 1.3. A bright nebula. The gas in the η Carinae Nebula emits light as a result of being heated to 10 000 K by the hot stars it contains. (See Appendix C for a listing of the Greek alphabet)

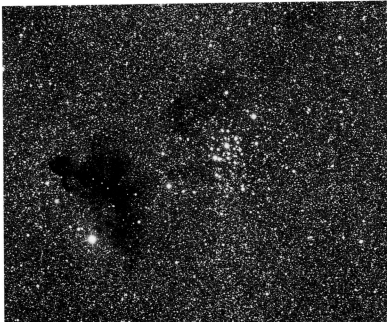

Figure 1.4. The patches of sky devoid of stars are dark nebulae, sometimes called dark clouds. The stars behind the clouds are blocked from our view by dust grains inside the clouds. The gases inside dark nebulae do not emit light since there are no suitable stars to heat them.

interstellar vacuum to the best laboratory vacuum one would face a density contrast greater than that between a breeze and a brick wall.

We can compare the gas in interstellar space with the gas in the Earth's atmosphere in another way. If a rocket were to take off from the surface of

Figure 1.5. The Milky Way as it appears to the eye (bottom), and as seen by the COBE infrared satellite (middle and top). The middle picture shows 1.2–3.4 μm radiation and gives the best impression of how stars are concentrated in a thin disk around the center of our Galaxy. At visible wavelengths (bottom) we see a more confusing picture, because the distant parts of the Galaxy are hidden from our view by thick dust clouds, some of which show up as dark patches near the center of the Galaxy. The top image shows the 25–60 μm infrared emission, which is produced almost entirely by warm dust particles in the interstellar medium.

the Earth and travel in a straight line across the Galaxy, it would intercept more gas in the first 10 km of its path through the Earth's atmosphere than on the whole of the rest of its journey to the edge of the known Universe. Interstellar matter may be very thin but the Universe is so vast that there is an awful lot of it. Our Galaxy contains some 10^{37} tons of interstellar gas, enough to make 10 billion stars the size of the Sun.

1.5 Interstellar matter, *pro* and *con*

To some astronomers, interstellar matter is a curse. It dims stars, distorts colors, conceals parts of our Galaxy and limits how far we can see into space. It prevents us from detecting some kinds of radiation at all.

To other astronomers, interstellar matter is a cornerstone of their science. It is the stuff out of which new stars are made, and into which they eventually fade. Interstellar matter permits astronomers to trace the rotation of the Milky Way, to monitor the creation rate of new elements, and to determine the mass of distant spiral galaxies. The most distant objects known – certain quasars – shine by the light of the gases they contain, while the youngest stars in the Galaxy are discovered by means of the radiation from the dust particles that surround them. Interstellar space provides the scientist with a laboratory whose vastness and emptiness allow atoms to act in ways that cannot be duplicated on Earth. Enormous gaseous reservoirs have been found containing chemical compounds never previously found on Earth. Cosmic masers exist that radiate more power than the Sun.

Perhaps most mysterious is the possibility that the Universe may be filled with material which none of our current telescopes can detect. There is a mounting suspicion that interstellar space contains vast quantities of **dark matter** which produce gravitational forces but emit neither light nor any other form of radiation. Dark matter deflects the paths of stars, binds clusters of galaxies together and may eventually cause the Universe to collapse in on itself. If the existence of dark matter is confirmed, some fundamental laws of physics may have to be rewritten. Understanding the nature of the interstellar matter that *is* detectable by telescopes is a necessary step toward the resolution of this mystery.

2 Light and radiation

Only within our own Solar System can astronomers have direct contact with the objects of their study. Almost everything else we know about the rest of the Universe is derived by studying the **radiation** emitted by distant matter. The word radiation encompasses benign forms of energy such as heat, light and radio waves as well as the more hazardous X- and gamma rays; in astronomy and physics the word radiation does not have the sinister associations it has acquired in this post-nuclear age.

One way of describing the progress of astronomy since prehistoric times is to consider the increasing sophistication by which radiation from space has been analyzed. Four great eras can be recognized.

In the **first** era, which lasted until the sixteenth century, the only radiation that could be studied was the light that directly entered the eyes of men and women as they gazed at the skies. Only bright objects like stars, planets and comets could be seen. No more could be known about a star than its apparent position in the sky, its brightness and its color.

The **second** great era is marked by the invention of the telescope, which led to the discovery of planetary satellites, nebulae, galaxies, stellar motions and brightness variations. Gravity became recognized as a major force in the Universe, but the nature of the stars and planets themselves remained a mystery.

The **third** great era started a century ago with the development of photography, spectroscopy and, later, electronic detectors to analyze the light received by the telescope. The astronomical observatory became an extension of the physics laboratory, and the precise measurements of quantities such as wavelength and brightness made it possible to apply sophisticated new theories such as quantum mechanics to explain how stars worked; the subject of astrophysics was born.

The **fourth** great era started in the 1940s and continues through the present. It is marked by the development of a vast array of new kinds of telescopes and instruments that are sensitive not just to light waves but to all the many other kinds of radiation that arrive at Earth from space, such as radio waves, X-rays and infrared radiation. Astronomers now have access to a far greater range of data about the heavens than was available to their predecessors. They sometimes use the metaphor of 'opening windows' to describe their access to this hitherto hidden information. Regions of the Universe that were previously totally invisible, including the diffuse interstellar medium, can now be examined in detail. The technological progress in astronomy of the last 40 years is mainly the result of advances in electronic techniques for converting faint signals into measurable electric currents, but the development of computers and satellites has also been crucial.

2.1 Electromagnetic waves

With very few exceptions all the radiation that astronomers collect to study the skies are examples of **electromagnetic waves**, or **electromagnetic radiation**. These waves can be thought of as rapidly changing magnetic and electric fields that carry signals and energy through space. Electromagnetic waves of all kinds travel freely through a vacuum at the speed of light. The most distinctive feature of an electromagnetic wave is its **wavelength**, which is the distance between successive 'crests' of a wave. In practical terms there are essentially no limits to the wavelength of an electromagnetic wave. It can be larger than the diameter of the Earth or smaller than the nucleus of an atom. The range of wavelengths is sometimes called the **electromagnetic spectrum**. Radiations of different wavelength, although fundamentally the same phenomenon, behave in such different ways that their underlying relation to each other may be not at all apparent; scientists therefore like to classify electromagnetic waves into several broad categories. These categories, which are listed here in order of decreasing wavelength, are partly historic, partly physiological and partly technological.

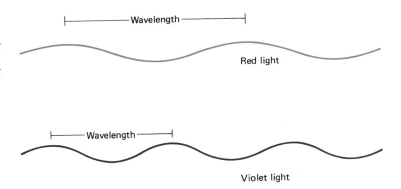

Figure 2.1. The wavelength of an electromagnetic wave is the distance between successive crests. Different wavelengths of visible light appear to us as possessing different colors. Violet light has the shortest wavelength; red light has the longest.

Radio waves comprise all electromagnetic waves longer than about 0.3 mm. They are easy to generate and detect, and form the basis for most long-distance human communication. Radio and television broadcasting is mostly done using wavelengths in the range of a few meters to a few hundred meters. Different radio stations and television channels avoid confusing each other by transmitting at different wavelengths which can be selected by 'tuning' a receiver. Short radio waves, less than about 30 cm in wavelength, are sometimes called **microwaves**; they are easier to control and focus than longer radio waves, so are often used for radar and for telephone communication. Most radio waves travel freely through the Earth's atmosphere and can pass right through solid objects such as a human body. Microwaves, on the other hand, tend to become absorbed in the first few centimeters of a solid or liquid – hence the use of microwave ovens to cook food. Most radio astronomy is done in the microwave region. Astronomers refer to the shortest radio waves – those between 0.3 mm and 3 mm wavelength – as **millimeter waves**.

Infrared waves have wavelengths between about 0.3 mm and 0.7 μm. They are invisible to human eyes, but in large quantities can produce the sensation of heat on our skin. All objects on planet Earth emit copious infrared radiation. The amount emitted depends on the temperature of the object. Specially designed infrared television cameras are therefore sometimes used to scan for temperature differences in an image of warm and cold objects. The differences may be subtle, as in medical diagnosis of blood flow by measurement of skin temperature, or dramatic, as in a heat-seeking guided missile.

Visible light waves are, by definition, those that can be detected by human eyes. They have wavelengths between 0.4 and 0.7 μm (4000 and 7000 Å); radiation outside of this range produces no reaction in the human retina. Light of different wavelengths is perceived by the eye as having different colors. The sequence of colors is that of the rainbow – red, orange, yellow, green, blue, violet – with red having the longest wavelength and violet the shortest. The precise range of wavelengths to which the eye is sensitive is the evolutionary result of several influences. Almost certainly, one significant factor is the Sun itself, which emits more power as visible light than as any other kind of radiation.

Ultraviolet waves are the waves immediately shorter than light waves, in the range 4000–100 Å. They are the waves that produce suntan, but in excessive strength they are damaging to eyes and skin. Their ability to destroy many kinds of molecules is exploited in sterilization processes. The ultraviolet waves in sunlight are a major cause of fading in colored objects.

X-rays have wavelengths between 100 Å and 0.1 Å (10^{-8} and 10^{-11} m). They can penetrate many solids and liquids that are opaque to visible light, making them extremely useful for medical diagnosis. Although some X-rays are produced by the Sun, we are shielded from them by the Earth's atmosphere, so that most of the X-rays we normally encounter are produced by human activity.

Gamma rays (also written 'γ-rays') are the shortest electromagnetic waves routinely studied by scientists, with wavelengths of less than 10^{-11} m. They are produced by nuclear reactions and are usually dangerous to human life. Some gamma radiation occurs naturally as the result of radioactivity in the Earth's rocks; the small genetic changes that this radiation produces in living cells are one of the factors controlling the rate of physiological evolution on Earth.

For studies beyond the Solar System there are only three types of data that astronomers collect which are non-electromagnetic in origin. These are **cosmic rays**, **neutrinos**, and **gravitational radiation**. Cosmic rays are discussed in Chapter 9; neutrinos and gravitational radiation do not provide any useful information about interstellar matter.

2.2 The speed of light

All kinds of electromagnetic radiation travel through empty space at the speed of light, namely 300 000 km s^{-1}. Radiation can be slowed as it travels through a gas or other transparent substance, but this slowing is usually negligible in interstellar space. The speed of light is so large that in everyday life we tend to think of signals travelling instantaneously; it takes less than a millionth of a second for light to cross a football field, and less than a fiftieth of a second for a radio wave to cross the Atlantic Ocean.

For astronomers, however, these delays can be substantial. Light takes eight minutes to travel the 150 million km from the Sun, four years to get here from the nearby star α Centauri, and 25 000 years for the journey from the center of the Milky Way Galaxy. The signals we receive from, and the images we perceive of these objects are therefore always out of date. For studies of interstellar matter within our Galaxy these delays make little difference to our overall view; 25 000 years is a very short period of time when compared with the age of the Galaxy, which is around 16 billion years, and to the time it takes the Sun to revolve once around it – about 240 million years.

2.3 Frequency

Another way of characterizing an electromagnetic wave is by its **frequency**. A wave passing a point in space produces an electrical signal that oscillates in strength, much as a buoy bobs up and down as an ocean wave passes beneath it. The number of times per second that the wave oscillates is called the frequency and is measured in cycles per second or, more commonly, 'hertz'. It takes less time for one cycle of a short-wavelength wave to pass a particular point in space than it does for a cycle of a long-wavelength wave. Consequently the shorter the wavelength of a wave the higher is its frequency. A radio signal of wavelength 3 m has a frequency of 10^8 hertz (Hz), or 100 megahertz (MHz) while a light signal of wavelength 0.5 μm has a frequency of 6×10^{14} Hz.

Waves in a near-vacuum can be described equally well by their wavelength or their frequency, and it is largely a matter of tradition which designation is more common. Light is usually defined by its wavelength while radio waves are commonly described by their frequency. The numbers on the tuning dial of a VHF or FM radio receiver (usually 88–108) refer to the frequency in megahertz of the signal being received. The mathematical relationship between wavelength and frequency is given in Appendix D, together with several other equations involving wavelength, temperature and energy.

2.4 Photons and energy

Electromagnetic radiation cannot be produced in infinitesimally small doses. A cornerstone of the **quantum theory** in physics is the idea that there is a minimum unit of electromagnetic energy that can exist at any particular wavelength. This unit is called a **photon**, and it is often useful to consider an electromagnetic wave, such as a light ray, as a stream of individual photons each of which behaves as a small bundle of waves. For some purposes it is convenient to think of a photon as a particle, but the analogy has to be used with caution because, unlike other kinds of particle, photons are always in motion at the speed of light and cannot exist at rest.

A single photon consists of waves of a single frequency, and it is this frequency which determines the amount of energy that the photon is carrying. The energy of a single photon in ergs is equal to its frequency in hertz multiplied by the number 6.6×10^{-27}. This number is called **Planck's constant**. It is one of the most fundamental constants of nature and is named after the German physicist Max Planck (1858–1947) who was a founder of quantum theory. An individual high frequency (i.e. short wavelength) photon carries more energy than a low frequency (long wavelength) photon. In other words, X-rays come in larger sized bundles of energy than radio waves. As a result, a pulse of radio waves with a total energy of one joule contains many more photons than a pulse of X-rays with the same total amount of energy. The link between energy and wavelength is so strong that X-rays and γ-rays are often described in terms of their photon energies rather than their wavelengths. A photon of 1 Å wavelength has an energy of 12 000 electron volts (eV), so may be described as a '12 keV X-ray'.

A single photon of light contains very little energy – about 2 eV (3×10^{-12}

ergs). Trillions of photons enter our eyes each second during daylight hours; the number is so large that we do not notice any flickering as the individual bundles of energy hit our retinas. Many of the instruments built by astronomers for use on their telescopes, however, are so sensitive that they can detect and count individual photons as they arrive from space.

2.5 The transport of energy

The link between photons and energy is very important in astrophysics, because electromagnetic waves provide the most important means of moving energy about the Universe. For example, the energy we receive from the Sun travels as visible, ultraviolet and infrared radiation from the Sun's surface to us. Electromagnetic radiation is one of the three major methods of energy (or heat) transport that we experience on Earth. The other common methods of heat transport, namely conduction and convection, are relatively less important in the Galaxy than they are on Earth; they depend on collisions occurring between neighboring atoms rather than on photons travelling through space. Even within the Sun, where collisions between atoms are much more frequent and violent than on the Earth, radiation is more important than conduction or convection for transporting energy.

2.6 Distortion of electromagnetic waves

If an electromagnetic wave travels through completely empty space it suffers no distortion except for a steady decrease in strength as its energy is spread out into larger and larger volumes. The fact that the wavelength and direction of travel of a photon may remain unchanged for millions of years is most fortunate for astronomers, since these well-preserved photons provide us with all we know about distant galaxies; the directions of travel of the waves allow us to put together an image; while their precise wavelengths tell us of the type of atoms in the galaxy, and of the speed at which it is moving. A beam of photons travelling through space constitutes an almost-perfectly preserved historic record of past events.

In space that contains interstellar matter, however, electromagnetic waves can become affected in various ways that depend on the nature of the medium and on the wavelength of the waves. Much of our knowledge of the interstellar medium comes from studying the various ways in which electromagnetic waves are distorted in space. The most important such effect is **absorption**, whereby part of the energy in the incoming wave is extracted into the interstellar matter. The energy may later be reradiated back into space with a different wavelength and in a different direction. A common example of absorption occurs when light from a star passes through a cloud containing dust grains. Part of the light is intercepted by the dust and its energy goes into raising the temperature of the dust grain slightly. Absorption causes a dimming of the radiation rather than a total blocking of it, and an interstellar cloud may be referred to as relatively 'transparent' or 'opaque' at some wavelength depending on the degree of dimming of the radiation passing through it. Some absorption processes apply only at certain discrete wavelengths, but the more powerful processes produce dimming over broad ranges of the electromagnetic spectrum. At visible and ultraviolet wavelengths dust grains do most of the absorbing, but at X-ray and radio wavelengths gaseous atoms are more important. Generally speaking, the more matter there is in a region of space the more radiation will be absorbed, but the transparency of a cloud to a particular type of radiation may also depend on its temperature and on how it is moving.

A second important way that radiation may be distorted in space is **scattering**. The direction of travel of a photon is altered, but its wavelength is unchanged. No energy is absorbed by the interstellar matter in this case. The most common astronomical example of scattering occurs in a reflection nebula, in which light from a star bounces off dust grains in the star's vicinity. To us as observers the light then appears to be emanating from the dust rather than the star.

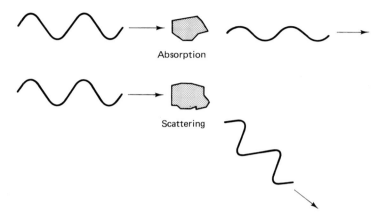

Figure 2.2. Absorption occurs when energy is removed from an electromagnetic wave and retained in the dust grains in the form of heat. Scattering occurs when part of the wave is diverted into another direction. Both processes may occur simultaneously.

Electromagnetic waves travel through matter more slowly than through a perfect vacuum, a phenomenon that gives rise to **refraction**. The amount of slowing is extremely small in interstellar space and has been detected only in a few special cases; the longest delays that have been measured, in the radio signals from pulsars, are only a few seconds in a journey that has taken the waves thousands of years. Refraction also can cause a change in direction, or bending of the wave. This effect is very small in interstellar space, but can cause some problems at long radio wavelengths.

2.7 Walls and windows

The fact that interstellar matter distorts electromagnetic waves is a mixed blessing. Small amounts of distortion make the matter visible, but too much absorption makes it opaque. At certain wavelengths and in certain directions the interstellar medium in the Milky Way Galaxy blocks almost all the radiation impinging on it, and constitutes an impenetrable wall. There are two main wavelength ranges where this blocking occurs. Radio waves with wavelengths longer than about 10 m are absorbed by free electrons, while ultraviolet waves with wavelengths of a few hundred ångströms are strongly absorbed by neutral hydrogen and helium atoms. The reasons why the interstellar medium is so opaque at these wavelengths are explored in later chapters in this book.

Astronomy is essentially impossible at wavelengths where the Galaxy is opaque. Luckily, these unusable wavelengths are not crucial for astronomers, who generally consider themselves fortunate that the Solar System lies in a part of the Galaxy from which only a small amount of information is excluded. A more pressing problem, however, is that much of the radiation which reaches the Solar System cannot penetrate the Earth's atmosphere and is absorbed by molecules and atoms within a few kilometers of the surface of the Earth. At wavelengths where the atmosphere is opaque, astronomers must use telescopes which have been placed outside the Earth's atmosphere. Until the Moon is colonized this can be done only by putting the telescope into a satellite in orbit around the Earth. Such a feat is expensive, inconvenient and puts severe limits on the size of the telescope. Consequently astronomy is done from the ground – albeit usually the tops of high mountains – wherever possible. Satellites, however, are always necessary for X-ray astronomy, and are needed for some kinds of infrared, and almost all kinds of ultraviolet observations.

Astronomers speak of **atmospheric windows** when referring to those ranges of wavelengths where ground-based telescopes can be used. The most important windows are at visible and radio wavelengths; a number of narrower ones occur in the infrared and ultraviolet regions. The transparency of many of these windows depends greatly on the weather conditions at the observatory site. Cloud cover can effectively block light from the visible window, while fluctuations in humidity are the enemy of infrared astronomy. Radio telescopes are the least affected by the atmosphere and can be located almost anywhere on Earth, but the geographical location of an infrared- or visible-wave observatory is crucial.

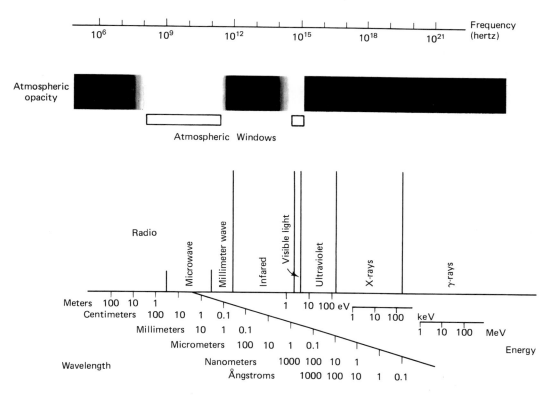

Figure 2.3. *The Earth's atmosphere is opaque to many kinds of electromagnetic radiation. Radio waves, light waves and some infrared wavelengths reach the ground, but astronomers must resort to telescopes carried in satellites if they want to study γ-rays, X-rays and ultraviolet waves reaching the Earth from the Galaxy.*

2.8 Tools of the trade: telescopes

This book is about what we know and what we do not know about interstellar matter. Sometimes our lack of knowledge is due to lack of imagination, but in many cases it is due to technical limitations in our ability to observe the sky. To appreciate these limitations we need some idea of the tools that are available to astronomers and of the problems that must be faced in using these tools. Most chapters in this book therefore contain a digression in which relevant types of astronomical techniques are briefly described. We start, inevitably, with the subject of telescopes.

Astronomers have used telescopes to study the sky ever since the time of Galileo. Nowadays they have telescopes for observing all kinds of electromagnetic radiation. As we will discuss in later chapters each range of wavelengths has its own problems and techniques, so that vastly different design principles are used for, say, radio and gamma ray telescopes. Even if we exclude the radical designs necessary to cope with these extremes of wavelength and confine ourselves to visible wavelengths we find that there have been enormous changes in the ways that astronomers have used telescopes over the past four centuries. The telescopes that present-day professional astronomers use differ from Galileo's in several fundamental ways:

(1) They are much larger. Galileo's telescope had a diameter of a few centimeters. Most present-day visible-wave astronomy is done using telescopes with diameters at least 100 times larger. The Hale 200-inch telescope at Mount Palomar in California – for many years the world's largest – has a diameter of 5 m. The Keck telescope in Hawaii is comprised of 36 closely spaced mirrors giving it an effective diameter of 10 m. Large diameters give the astronomer the ability to gather more light, particularly from objects so faint that only a few photons reach the telescope each second.

(2) They use mirrors, not lenses. Curved mirrors collect and focus light much as lenses do, but have the advantage that only one expensive polished surface is necessary instead of two or more. Other advantages are that a mirror can be supported from behind by a steel framework whereas a lens can only be supported by its edge. Also, a mirror focuses all wavelengths of light to one place, whereas a lens, unless very carefully designed, focuses red and blue light differently.

(3) They are designed for electronic instruments, not for human eyes. Television cameras are more sensitive and can operate over a wider range of wavelengths than can human eyes, so astronomers nowadays almost always view the sky through a video monitor rather than an eyepiece. Photography is still sometimes used, but for most observations the light collected by the telescope is focused into some kind of electronic instrument where it is turned into signals that are stored for later examination and analysis using a computer.

(4) They produce better images. A good modern telescope can produce images which show details smaller than 1" across – about 100 times finer than can be seen by the unaided human eye. The main limitation in optical quality is usually the Earth's atmosphere; the same kinds of atmospheric fluctuations that cause a star to twinkle can blur the image of a galaxy in a large telescope. The effect can be reduced by a careful choice of where to build the telescope, but the problem of atmospheric blurring can be solved completely only by placing it in orbit above the Earth's atmosphere. For certain kinds

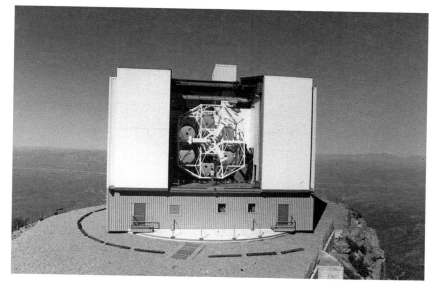

Figure 2.4. The Multi-Mirror Telescope (MMT) in Arizona is an example of a new trend in telescope building, in which several comparatively small mirrors are used to focus light to a single point, rather than one very large expensive one. The MMT has six mirrors of 1.8 m diameter each which together are equivalent in area to a single mirror 4.4 m across.

of astronomy the resultant improvements in image quality are worth the extra cost involved. The Hubble Space Telescope, launched into orbit by NASA in 1990, was the first large telescope designed to obtain pictures of the skies unblemished by the Earth's atmosphere. Unfortunately technical problems have so far prevented the telescope from performing at its full potential.

A large modern telescope costs tens of millions of dollars to build, and several million dollars a year to operate. There are very few institutions which can afford these costs by themselves. Many observatories are therefore operated cooperatively, with astronomers from different organizations, or even countries, sharing the available nights between them. International cooperation in astronomy is as much a consequence of geography as economics or politics. To be operated efficiently a telescope must be located where the weather is good, the skies are dark and unpolluted, and the air is clear and smooth. Such conditions are found on only a handful of high mountain peaks throughout the world. Most major telescope construction in the last decade has taken place in one of four locations; Hawaii, Arizona, Chile or the Canary Islands, and this trend is likely to continue in the next decade.

3 Atoms and spectra

In everyday life we spend most of our time looking at solid objects. What our eyes usually see is sunlight that has been scattered off the surfaces of the objects in our view. Our eyes form images of the world that convey patterns of shape, color and motion to our brains, but it is our accumulated experience of how light scatters off surfaces of different materials at different angles that allows us to identify and understand the world in which we live. There are some exceptions. Luminous objects, such as the Sun, an electric light or a flame, are seen by the light which they themselves generate, while a transparent window may be perceived by the effect it has on light passing through it. On the whole, though, our view of the world in which we live is a view based on scattered light.

In astronomy, scattered light is the exception. There are only two kinds of common celestial phenomena that depend on scattered light for their appearance. First we have Solar System objects such as the Moon, the planets, and the comets. They give off no light of their own, and shine by reflecting the Sun's light back toward us; if we could switch off the Sun, Venus, Mars, and the other planets would almost immediately disappear from our view. The second phenomenon that depends on scattering is the **reflection nebula**. There are certain places in the Galaxy where a cloud of interstellar dust grains lies close to a bright star. When light from the star scatters off the particles in the cloud, the dust appears to glow, like tobacco smoke in a flashlight beam.

Figure 3.1. Reflection nebulae surrounding stars in the Pleiades cluster. Light from the bright stars scatters off the interstellar dust grains, making it appear that the dust itself is emitting light.

Away from the Solar System and away from reflection nebulae, scattered light does not play a major role in astronomy. One reason is that most of the Universe consists of gases, which tend to be much more transparent than solids or liquids; another is that most of the Universe is very dark, with insufficient light to illuminate any solid objects that may be around. Most of the radiation that astronomers study is *not* scattered; we therefore cannot rely on our everyday experience for interpreting the heavens. We need to look in much more detail at the ways that radiation can be generated and the ways that it can be affected by matter. To understand the links between radiation and matter we must borrow a number of ideas from physics. Fortunately, because most of interstellar matter exists as such tiny units, the physics is comparatively simple. At its root is the nature of the atom itself.

3.1 The structure of the atom

At the core of every atom is its nucleus. The nucleus contains most of the mass of an atom, but occupies only a very small fraction of its volume. It contains a number of identical protons, each of which has a positive electrical charge. The most fundamental property of an atom is the number of protons it contains, or its **atomic number.** Atoms of different atomic number are referred to as different chemical **elements**. There are only 92 naturally occurring elements in the Universe. The simplest, with atomic number 1, is hydrogen, the most complicated is uranium at 92. The atomic number of an atom determines whether it will be a solid or a gas, how it will react with other atoms, and even what color it will be. The most common elements, together with any others which play an important role in the interstellar medium, are listed in Appendix E. As we shall see, elements with small atomic numbers are much more plentiful than those with large atomic numbers. Astronomers sometimes refer to elements 1–5 as the **light elements** and elements 6–92 as the **heavy elements**.

With the important exception of hydrogen, all atomic nuclei contain a number of neutrons. A neutron has about the same mass as a proton but is electrically neutral. The neutrons affect the mass of the nucleus but not much else. Generally the number of neutrons in an atom is similar to the number of protons, but different atoms of the same element can have different numbers of neutrons; the varieties are referred to as **isotopes**. The total number of protons plus neutrons is called the **atomic weight**. For most elements there is one isotope that is much more common than the others. For example 99% of the carbon atoms on Earth have six neutrons in their nuclei; the remaining 1% have either seven or eight neutrons.

Surrounding the nucleus of an atom is a cloud of electrons. An electron has only 1/1836 of the mass of a proton or neutron and so contributes negligibly to the total mass of an atom. Each electron, however, carries a negative electrical charge that exactly balances that of the proton. The electrostatic force between the oppositely charged protons and electrons helps hold the atom together. Under normal circumstances atoms are electrically neutral and so contain equal numbers of protons and electrons. Under provocation, however, atoms may temporarily lose an electron and become **ionized**. When this happens the atom acquires a positive electrical charge equal to the number of electrons it has lost, and is referred to as an **ion**. In this book ions are indicated by one or more '+' suffixes; singly ionized oxygen is written O^+, while triply ionized sulfur is S^{+++}, or S^{3+}. The most common ion in astrophysics is the hydrogen ion, which consists simply of a single proton. Ions have a strong tendency to recombine with any free electrons they can find, so are fairly rare in the dense environment

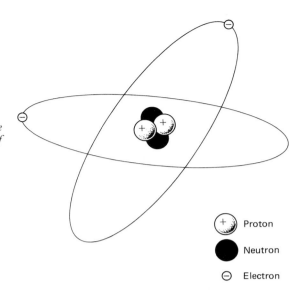

Figure 3.2. The neutral helium atom contains two protons and two neutrons in its nucleus. There are two electrons in orbit around the nucleus. This diagram is not to scale; the diameter of the electron orbits is 10^5 times that of the nucleus, so that the nucleus occupies only 10^{-15} of the atom's volume. The atom can become singly ionized by losing one electron or doubly ionized by losing both.

of the Earth. Where gases are thinner or hotter, ions have a better chance of surviving. Above about 100 km altitude most of the Earth's atmosphere is ionized. In interstellar space, where the densities are even lower, ions are very common. The gases in most bright nebulae, including the η Carinae Nebula (Figure 1.3), are essentially completely ionized.

3.2 Energy levels and transitions

The motions of the electrons in an atom are governed by the laws of **quantum mechanics**, which differ from those of classical physics in some major ways. One of the most important consequences of quantum mechanics is that the electrons in an atom can exist only in certain configurations, called **quantum states**. The quantum states of an atom are sometimes pictured as orbits of the electron around the nucleus. Quantum states differ from each other in the shapes of their electron orbits and in the average separation of the electron from the nucleus. The electrostatic force pulling the electron toward the nucleus depends on their separation, being stronger when the electron and nucleus are close together. Consequently, in some quantum states the electron is more tightly bound to the nucleus than in others, and a greater amount of energy is required to separate, or ionize, them. This leads to the idea of **energy levels**, which are groups of quantum states for which about the same amount of energy is required for ionization.

The quantum state in which the electrons are most tightly bound to the nucleus is of particular importance and is called the **ground state**. This is the state that an atom reverts to if left alone. States other than the ground state can only be reached if energy is added to the atom by some means. These states of higher energy are therefore called **excited states**. Figure 3.3 represents the energy levels of a hydrogen atom; these have been determined both by laboratory experiments and by solving quantum mechanical equations. By convention, the different energy levels in hydrogen are distinguished by different values of the letter n. The energy levels in Figure 3.3 may be visualized as a ladder. The lowest energy level (called the $n=1$ level) contains just two quantum states, including the ground state. The next highest ($n=2$) has eight quantum states at an energy of 10.2 eV above the ground energy level. Energy levels higher than $n=2$ become progressively closer to each other, and crowd together just below the limit of 13.6 eV. This limit is called the **ionization potential**. It is the amount of energy needed to ionize an atom out of its ground state. The ionization potential plays a very important role in the physics of the interstellar medium. It is listed for a number of common elements in Appendix E.

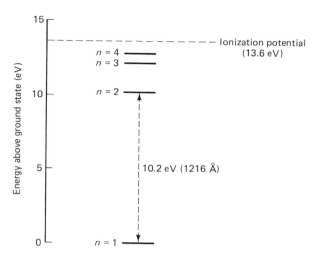

Figure 3.3. Energy levels of hydrogen. An undisturbed atom is usually found in the ground state, also called the $n=1$ state. When it is excited it temporarily jumps into a level with higher energy. Diagrams like this enable spectroscopists to calculate the wavelengths of photons emitted when atoms make transitions from one energy level to another. Note that the term 'ground state' strictly refers only to the lower of the two quantum states that comprise the $n=1$ energy level. See also Figure 4.1.

Under the right circumstances an electron may move from one quantum state to another. Such a move is called an **electronic transition**. Unless the initial and final quantum states are at exactly the same energy level, a transition will involve either the absorption or the emission of energy. Transitions from a lower to a higher value of n, which require the input of energy into the atom, are referred to as **upward transitions** while those that involve a decrease in n and the release of energy are referred to as **downward transitions**.

Transitions between energy levels can occur for a number of reasons. A **spontaneous transition** is a downward transition in which the atom, with no external provocation, moves from a high energy state to a lower energy state emitting a photon. The wavelength of the photon is determined precisely by the difference in energy between the upper and lower energy states; for example, the transition from the $n=2$ level to $n=1$ level in hydrogen produces an ultraviolet photon of wavelength 1216 Å.

Spontaneous transitions cannot occur upwards from a low energy level to a higher one, because of the need to supply energy to the atom. There are two ways of exciting upward transitions. One way is by a **stimulated transition**, whereby the atom absorbs a photon that has a wavelength corresponding precisely to the energy difference between the two levels. A stimulated transition can only occur if there are photons of a suitable wavelength passing in the vicinity of the atom. A second means of exciting an atom is by a **collisional transition**, in which the required energy is supplied by the impact of a colliding particle. Collisional transitions are more likely to occur in dense gases, where collisions are more frequent. Stimulated and collisional transitions are also possible in the downward direction; an atom in an excited state may be induced to drop to a lower energy level by the influence of photons of just the right energy or by a collision with another particle.

Because the neutral hydrogen atom contains only one electron, its pattern of energy levels is particularly simple. Other atoms are more complicated because of the interactions among the several electrons surrounding the nucleus. Each chemical element has a different set of quantum states and energy levels, and therefore emits and absorbs distinctive wavelength photons. There are differences between the energy levels of different isotopes of the same element. Also, an atom which is ionized has quite different energy levels from the same atom when it is neutral. These varied patterns of energy levels make it possible to identify atoms and ions from the wavelengths of the radiation they emit and absorb. This principle is one of the cornerstones of the science of **spectroscopy**, which is the analysis of radiation according to its wavelength. Underlying all spectroscopy is the discovery that atoms of the same chemical element are identical everywhere in the Universe. The relative proportions of the 92 different elements and their isotopes may vary from place to place, but throughout the Universe all atoms with the same mix of protons, neutrons and electrons are identical and emit and absorb photons of exactly the same wavelengths.

3.3 Spectral lines

When light is passed through a prism it becomes sorted according to the wavelengths of its photons. The resulting pattern of light as a function of wavelength is called a **spectrum**. In a physics laboratory a spectrum may be viewed through the eyepiece of a spectroscope or projected as a colored band of light on a screen, but in astronomy it is usually displayed either as a graph of brightness against wavelength (Figure 3.4) or as a pattern on a photographic plate (e.g. Figure 3.5). Wavelengths that stand out as being particularly bright or particularly dim are referred to as **spectral lines** because of the way they appear as small streaks on a photographic spectrum. In principle, spectrum lines can occur at any wavelength, not just in the visible part of the spectrum. Under most conditions, however, the strongest spectrum lines from atoms and ions are found at ultraviolet and visible wavelengths.

Different astronomical objects produce quite different spectra. A hot gas produces a spectrum that contains **emission lines**. Atoms become excited out of their ground states by collisions with other atoms, then undergo spontaneous downward transitions to the ground state either directly or through a series of transitions using intermediate energy levels. Each downward transition produces a photon at one of a number of a specific wavelengths. Figure 3.4 shows the spectrum from hot gas in a bright nebula. It consists almost entirely of emission lines.

For a gas to produce an emission line spectrum it must have a low density; if the atoms in the gas are forced too close together their quantum states become distorted and their energy levels change in complicated ways. As a result, the atomic transitions no longer have precise wavelengths and the spectrum lines become **broadened**. In the case of extremely high density the emission lines become so broadened that they blend together to produce a **continuous spectrum**. Such conditions are found in solid bodies, for example. In an object that is producing **continuum radiation** emission lines are not seen and it is no longer possible to identify photons as having been produced by particular elements. On Earth, we can see examples of both emission line spectra and continuous spectra in commercial electric lighting. An ordinary domestic light bulb contains solid tungsten wire that becomes white-hot when current is passed through it. Hence it emits a continuous spectrum. Fluorescent tubes, neon advertising signs, and orange-colored sodium street lights, on the other hand, produce their light via low-density gases. They produce an emission line spectrum. We will return to

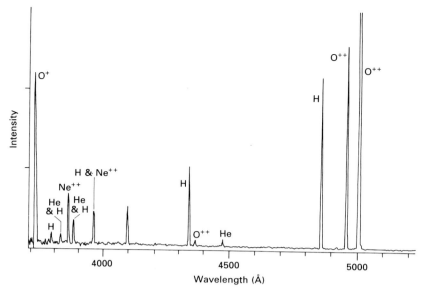

Figure 3.4. Spectrum of hot gas (~ 10 000 K) in a bright nebula. In this electronically-recorded spectrum an emission line shows up as a spike in this graph of intensity against wavelength. The most prominent features here are emission lines from atoms or ions of hydrogen (H), helium (He), oxygen (O) and neon (Ne).

continuous emission, and how it relates to the temperature of an object, in Section 3.8.

The density of the hot gas in the Sun and other stars is high enough that, to a first approximation, its light has a continuous spectrum. On closer examination the spectrum is seen to be crossed by a number of dark bands called **absorption lines** (Figure 3.5). These lines arise when a light with a continuous spectrum passes through a region of comparatively cool gas –

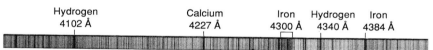

Figure 3.5. Absorption line spectrum of the Sun recorded on a photographic plate. The Sun emits strong continuous radiation, but at certain wavelengths the light is absorbed by cooler atoms in the Sun's upper atmosphere. Absorption lines show up as vertical streaks on this picture. Identifications and wavelengths are given for some of the more prominent lines.

such as the upper layers of the Sun's atmosphere. Upward stimulated transitions occur in the cooler gas as the atoms are excited to higher energy levels by the photons that pass through it. Only photons with wavelengths corresponding to specific energy level differences can be absorbed. As the excited atoms spontaneously revert back to their lower energy levels some photons with the same wavelengths as the absorbed photons will be produced. These regenerated photons, however, are emitted in all directions indiscriminately and will generally not follow the same path as the original light. Consequently, after passage through the cool gas, the continuous spectrum appears darker at specific wavelengths.

The analysis of spectrum lines can become extremely complicated; the same body of gas can simultaneously show some wavelengths as emission lines and some as absorption lines. The identical spectrum line can appear in emission in one direction and absorption in another. However the general rule is that a body of gas shows emission lines when it is looked at in isolation, but will produce absorption lines in the continuous spectrum of an object behind it if the temperature of the background object is higher than the temperature of the gas.

Figure 3.6. Hot diffuse gases produce narrow emission lines. As the gas density increases the lines become broadened. Eventually, when the density approaches that of a solid the emission lines all blur together producing a continuous spectrum. Absorption lines are seen when a source of bright continuum radiation is viewed through a region containing a cooler diffuse gas.

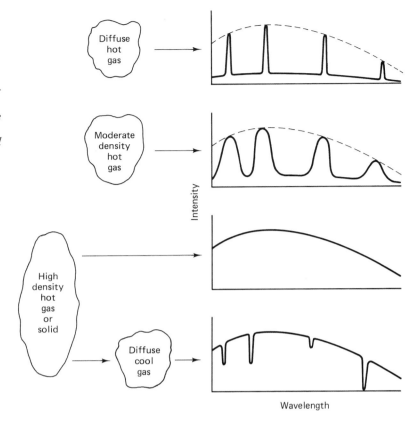

3.4 The Doppler effect

Spectrum lines provide information not only about the identities and concentration of atoms; they also tell us how they are moving, by means of the **Doppler effect**. Photons emitted from an object moving toward us appear to have a shorter-than-normal wavelength; photons emitted from an object moving away from us appear to have a longer-than-normal wavelength. These phenomena are referred to as the **blueshift** and **redshift** respectively, since blue and red (or more correctly violet and red) are at the extreme shortwave and longwave ends of the visible spectrum. The expressions 'blueshift' and 'redshift' are something of misnomers because except for some exceedingly distant galaxies, the changes in wavelength due to the Doppler effect are much too small to make perceptible changes to the color of an object. What *can* be measured, though, is the small difference in wavelength between a particular spectrum line as seen in a distant moving star or gas cloud and as measured in a stationary lamp on Earth. From this difference in wavelength it is easy for an astronomer to calculate the speed at which the star or cloud is moving towards us or away from us (see Appendix F for details). This speed is usually called the **radial velocity** of the object. Sideways motions do not cause measurable Doppler effects, and cannot be determined from a spectrum.

The Doppler effect has all sorts of hidden advantages which make it one of the most useful tools that astronomers have for analyzing their data. As we shall describe in later chapters of this book we can use the Doppler effect to estimate the distance to an interstellar cloud, to distinguish between clouds that lie in front of or behind each other, to measure the temperature of a cloud, to find out whether a cloud is about to turn itself into a star, to measure the shape of the Milky Way Galaxy. Almost all of cosmology – the study of the origin and fate of the Universe – is based on measurements of the Doppler effect.

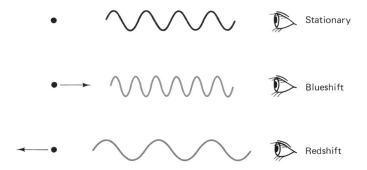

Figure 3.7. Photons emitted by an object travelling towards the observer have their wavelengths shortened by the Doppler effect; those from an object moving away from the observer have them lengthened.

3.5 Tools of the trade: spectroscopy

Spectroscopy is so powerful that it is used for about half of all astronomical observations. Spectra are obtained using specially designed instruments called **spectrographs**. There are many different designs, but the principle is always the same; to sort out light from a star or nebula according to its wavelengths and then record the information on a photographic plate or a computer. The sorting is usually done with a **diffraction grating**, which is a mirror covered with minute parallel straight grooves which reflect different wavelength light in different directions. For most purposes it is more powerful and more versatile than a prism. Other spectrographs are built around more complicated systems called **interferometers**, in which a system of moving mirrors is used to analyze the light.

In a simple spectrograph the telescope functions simply as a means of collecting light from one point in space, such as a star. More elaborate instruments can simultaneously record separate spectra for dozens of closely-spaced points. Such instruments are used for obtaining redshifts of many galaxies in a single cluster, or for studying how the gases are moving in different parts of a nebula.

Astronomical spectrographs have to be designed for their purpose. To identify chemical elements and measure redshifts of faint galaxies we need to cover a wide range in wavelength, and to be very careful about not wasting light; we do not need to measure wavelengths very accurately, however. For some other measurements the **spectral resolution**, which is a measure of ability of the spectrograph to discriminate between photons of slightly different wavelength, is important. We need to measure wavelengths to a precision of at least 1 part in 100 000 if we want to study gas motions in an interstellar nebula. The highest resolution yet achieved is in an instrument that searches for evidence of planets around nearby stars. It can measure wavelengths to an accuracy of better than 1 part in 20 million; such precision is sufficient to detect the Doppler shift of an object moving at only 50 km per hour – less than one thousandth the speed of the Earth's motion around the Sun. Spectrographs that work at very high resolution have their drawbacks; they tend to be very large – as big as the telescope itself in some cases – and they can only be used for bright stars that emit plenty of photons.

Spectroscopy can be performed on electromagnetic radiation of any type, though the technology involved varies greatly. Radioastronomical spectroscopy, for example, involves tuning radio receivers to the frequencies of interest rather than separating the radiation with a prism. This process is quite similar to selecting a particular broadcast by tuning a radio dial, except that radio astronomy receivers usually work by first collecting signals over a broad band of wavelengths, and then subsequently analyzing them into a large number of separately-tuned channels.

3.6 Molecules

In some parts of the Universe, including the Earth and its lower atmosphere, individual atoms are rare, and **molecules** are the norm. Molecules are small groups of atoms which are bound together and which share each other's electrons. The electrostatic interactions between the various electrons and nuclei in a molecule hold it together in a stable unit. Most of the gases we deal with in our daily lives exist in the form of molecules rather than single atoms. The most common atmospheric gases, nitrogen and oxygen, are found as N_2 and O_2 – simple molecules consisting of two identical atoms. Other common molecular gases, such as carbon dioxide (CO_2), ammonia (NH_3) and methane (CH_4), contain more than one type of atom. The only common gases found as single atoms on Earth are helium (He), which is used to inflate balloons: neon (Ne), which is used for decorative lighting: and argon (Ar), which fills 1% of our atmosphere but is practically useless. Hydrogen gas does not occur naturally on Earth, but exists in the molecular form H_2 when prepared industrially.

In interstellar space much of the gas occurs in the form of isolated atoms rather than as molecules. The reason for this difference is mainly the much lower particle density in interstellar space as compared with the Earth's atmosphere. When two oxygen atoms combine to form an oxygen molecule (O_2) energy is released. The molecule is stable because it cannot be separated unless that energy is resupplied by a suitable photon or by an energetic collision. What few separations or '**dissociations**' of oxygen molecules that do take place in the Earth's lower atmosphere are soon nullified because each free atom combines with another free atom at the first possible opportunity. Since air molecules or atoms collide with each other every few billionths of a second, such opportunities come fast and the balance between dissociation and recombination is tilted heavily toward the molecular state. In interstellar space, collisions between atoms are separated by years rather than fractions of a second, so that the opportunities for atoms to combine into molecules are much rarer. In the 1970s, however, large concentrations of interstellar matter were discovered in which the gases are almost entirely in the form of molecules. The gas in these so-called **molecular clouds** is much denser than normal interstellar gas, and is partly protected from dissociation by layers of dust particles that prevent harmful photons destroying the molecules.

A molecule has totally different quantum states and energy levels from its constituent atoms. Each type of molecule produces its own set of spectral lines and can be identified by them. The number of possible transitions for a molecule is much greater than for an atom because its energy levels are affected by its rotations and its vibrations as well as by the shape of its electron clouds. Molecules, like atoms, produce their strongest transitions in the ultraviolet, but, unlike atoms, also produce strong lines at infrared and millimeter wavelengths as they adjust their rotation rate or their intensity of vibration. The great richness of molecular spectra has been a major factor in recent decisions to build new telescopes for collecting infrared and millimeter radiation.

3.7 Solids

Interstellar atoms can also bind together to form pieces of solid material. These are usually referred to as **dust grains**. In a solid, as in a molecule, the binding arises through interactions among the electron clouds of the atoms. In a solid the atoms occupy fixed positions with respect to each other. Usually, these fixed positions are in regular patterns forming a crystal, but amorphous solids are known in which the atoms occupy irregular positions. Ordinary salt is an example of a crystal; window glass is an example of an amorphous solid.

Interstellar solids are more difficult to study than interstellar gases. Dust grains absorb and emit radiation over wide ranges of infrared, visible and ultraviolet wavelengths. They produce no sharp spectral lines by which they can be clearly recognized. Broad absorption and emission features occur in some materials and are used by planetary astronomers to explore the composition of the surfaces of planets and asteroids. They are too few and too imprecise, however, to provide unambiguous identifications of cosmic dust grains. As a result, the composition of interstellar dust particles is still a question of great controversy.

As far as we know there are no liquids in interstellar space. Interstellar dust grains grow by condensation directly from interstellar gas, and shrink by evaporation or disintegration. Until about 1983 the distinction between the gaseous and solid phases of the interstellar medium was quite simple. The largest known molecules known had only a handful of atoms, while the smallest dust particles had hundreds of thousands. There is now evidence for the existence of a group of molecules in interstellar space called **polycyclic aromatic hydrocarbons** (PAHs). Each of these molecules contains 50 or more atoms and behaves with some of the characteristics of a minute dust particle. Interstellar matter does not always fit into the neat categories scientists have devised to describe the Earth and its environment.

3.8 Temperature

We have used the word temperature several times already in this book. In daily life we usually judge temperatures by our sense of touch. Since tactile response is of little value when it comes to studying interstellar matter, we need to pause briefly and examine what physicists and astronomers mean when they use the word, and how they apply the concept of temperature to phenomena that are beyond our everyday experience. There are three aspects to temperature that we need to understand, namely its connection with random particle motions, its connection with electromagnetic radiation and its connection with spectral lines.

All matter is in a state of constant movement. The energy of that motion is called **kinetic energy**. When heat or another form of energy is added to an object the motions of its atoms become more violent. The nature of the motions depends on what the object is made from. If the object is an atomic gas the added heat makes the atoms move faster in all directions, colliding with each other and with their surroundings. The forces that these fast moving atoms exert on their surroundings is the explanation for **gas pressure**. If the heated object is a solid its atoms maintain their positions in the crystal but increase their jiggling motions. If it is a molecular gas then some of the heat energy goes into making the molecules move faster through space, and some into making them spin and vibrate more energetically. Because of their associations with heat these motions are all described as **thermal motions**. Temperature is a way of measuring the energy of thermal motions; some details are given in Appendix G. Temperatures in astronomy are always measured in the Kelvin or absolute scale, in which the temperature is directly proportional to the thermal energy. A temperature of

0 K (also known as **absolute zero**) corresponds to a total absence of thermal energy, a state that can be closely approached, but never reached. Other examples of the Kelvin temperature scale are given in Appendix B.

Another important property of temperature for an astronomer is that it is the main factor governing how heated objects emit electromagnetic radiation. Earlier in this chapter we saw how continuous radiation is produced when a solid or a dense gas is heated. The power that is emitted as a result of the thermal motions of hot matter is called **thermal radiation**. Thermal radiation from all objects is quite similar except that the mix of different wavelength photons changes with temperature. If a solid body is heated, the average wavelength of the photons it emits becomes shorter as the temperature rises. Objects at room temperature emit infrared radiation; at higher temperatures they start to emit some red light in addition to the infrared. As they become yet hotter, increasing amounts of progressively shorter wavelength visible radiation are radiated so that the object eventually becomes white hot. The precise form of the continuous spectrum depends somewhat on the object's size and composition, but it always bears some resemblance to an idealized theoretical curve which physicists call the **black-body** spectrum (Figure 3.8 and Appendix H). This is the spectrum that would be produced by an object that is perfectly 'black', in the sense that it is able to absorb all radiation that lands on it. The most important feature of black-body radiation is that for any temperature of the radiating body there is a particular wavelength at which the maximum amount of power is given off. This maximum wavelength depends strictly on temperature according to **Wien's law**; the wavelength of peak emission is inversely proportional to the absolute temperature. Wien's law can be applied to various astronomical phenomena. It confirms that the Sun, with a surface temperature of 5800 K, radiates most power at about 5000 Å, in the visible wavelength range. Stars that are much hotter than the Sun produce mainly ultraviolet energy. The Earth and its inhabitants, with a temperature of around 300 K, radiate most strongly at around 10 μm, in the mid-infrared.

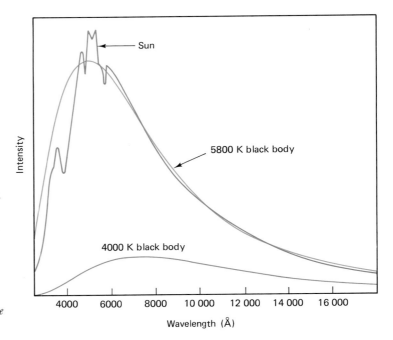

Figure 3.8. Thermal radiation from heated objects. The emission peaks at a wavelength which is shorter for hot things than for cooler ones. The curves are shown for theoretically perfect sources of radiation called 'black bodies'. The actual emission from the Sun is shown for comparison with the perfect black bodies.

A third property of temperature that we shall meet in this book is the way it can be used to predict the relative numbers of gas atoms or molecules in different energy levels. These calculations are important because they are the basis for interpreting spectral lines. These three ways of using temperature are closely linked via the branch of physics known as **statistical mechanics**.

As we shall find as we explore the different states of interstellar matter, the temperature of any particular location in space can often be impossible to define uniquely. The gas, the dust, and the radiation field may all have different temperatures associated with them. To some extent, it is these differences of temperature that keep interstellar space interesting, because without them, and the movements of energy they cause, the Universe would soon revert to a drab monotony.

4 Atomic gas in the interstellar medium

Hydrogen is by far the most common element in the interstellar medium as well as in the Universe. Describing the state of the hydrogen is therefore tantamount to describing the overall condition of the gas itself. Interstellar hydrogen can exist in three forms: molecular (H_2), atomic (H^0) and ionized (H^+). The idea that interstellar gas can exist in more than one distinct form has given rise to the idea of interstellar **phases**. The term 'phase' is used by physicists to refer to different forms of the same kind of matter, such as the solid, liquid and gaseous phases of a chemical. In the interstellar medium the balance between the molecular, atomic and ionized phases depends on the density, on the temperature and on the amount of ultraviolet radiation passing through the gas. For reasons which will emerge later, at almost every location in space one of these phases dominates over the others. It is therefore convenient to describe the interstellar medium as being divided into **H_2 regions**, **H^0 regions** and **H^+ regions**, depending on whether the hydrogen there is predominantly molecular, atomic, or ionized. All three kinds of region occur extensively in the Galaxy and will make their appearance in this book. In this chapter we will discuss interstellar atomic hydrogen (H^0), which is sometimes referred to by astronomers as 'neutral hydrogen'. Because it has a unique spectral line at a wavelength of 21 cm, atomic hydrogen is the easiest phase to observe directly. Note that in many astronomy books atomic hydrogen is referred to as **HI** and ionized hydrogen is referred to as **HII** where I and II are the roman numerals one and two.

4.1 The 21-cm line

Tens of thousands of different spectral lines have been observed by astronomers, but the **21-cm line** stands out as unique both in terms of its physics and of its importance for studies of interstellar matter. The lowest energy level of a hydrogen atom contains two quantum states which are slightly different. The difference is due to the fact that electrons and protons each behave in some ways like small magnets. Magnets exert forces on each other which depend on where they point as well as how far apart they are. In the case of a hydrogen atom the laws of quantum mechanics permit the magnets to line up in only two ways; pointing in the same direction ('parallel'), and pointing in opposite directions ('antiparallel'). The parallel state lies at a slightly higher energy than the antiparallel state; the small energy difference between these two states, 6×10^{-6} eV, corresponds to a photon of wave-

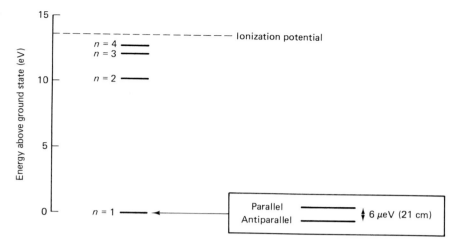

Figure 4.1. The lowest energy level of the hydrogen atom is split into two quantum states with slightly different energy levels. The energy difference between the levels corresponds to a photon of wavelength 21 cm. The split arises from a magnetic interaction between the electron and the proton in the atom. Splitting also occurs in the upper ($n \geq 2$) energy levels, but the corresponding spectral lines are too weak to observe.

length 21 cm. Transitions between the two energy levels can give rise to a spectrum line that can be studied with a radio telescope.

The 21-cm line is arguably the single most important spectral line in astronomy. It is the *only* spectral line that can be emitted by atomic hydrogen under normal interstellar conditions. All hydrogen spectral lines other than the 21-cm line have to make use of at least one energy level with $n \geq 2$. To excite a hydrogen atom into any of these excited energy levels requires *at least* 10.2 eV per atom – a million times more energy than the 21-cm line needs – and far more than is readily available in an H^0 region.

4.2 Atomic hydrogen in galaxies

A general rule in astronomy is that the closer an object is to us, the easier it is to study. Atomic interstellar hydrogen is an exception to this rule; for reasons that should become clear later in this chapter it is far easier to make sense of 21-cm maps of other galaxies than of the Milky Way Galaxy.

The galaxy Messier 83 lies about 5 Mpc distant in the southern hemisphere constellation of Hydra. If we could view the Milky Way from the depth of intergalactic space it would probably look rather like Messier 83. Figure 4.2 shows a comparison between an optical image of the galaxy and a map made by radio astronomers of its 21-cm emission. The light we see in the optical image comes almost entirely by the stars in the galaxy; the strength of the 21-cm line is a good indicator of where the atomic hydrogen is to be found. We can draw two important conclusions from a comparison of these pictures. First, the atomic hydrogen is spread out all over the disk of the galaxy, extending well beyond the bright visible regions of the galaxy. Second, there is more atomic hydrogen in the spiral arms than between them.

Additional valuable information comes from the Doppler shifts of the 21-cm line, which have been measured for all positions in the galaxy. Because the galaxy as a whole is moving away from us all the hydrogen has a redshift. When we look at the Doppler shifts in more detail we find that different parts of the galaxy are moving away from us at different speeds. The bottom right part of the galaxy is moving away from us faster than the top left because the galaxy is rotating and because it is tilted toward us. The rotation is clockwise as we view it, and the sense of the tilt is that the top right part of the galaxy is closer to us than the bottom left. By measuring the Doppler shifts in different parts of the galaxy we find that it does not all rotate at the same speed. A star in the outer regions takes about 160

Figure 4.2. Atomic hydrogen in the galaxy Messier 83. A map of the 21-cm radio emission (top) shows both similarities to and differences from an optical image (bottom). Measurements of the Doppler shifts of the 21-cm line allow us to separate gas that is rotating away from us (colored red) from gas that is rotating towards us (colored blue).

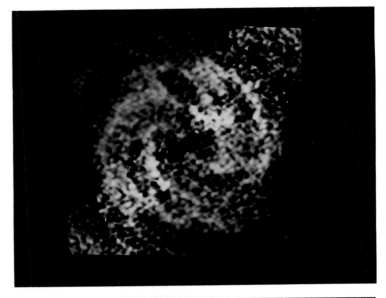

million years to make one orbit, but stars closer to the center of the galaxy take less time. This **differential rotation** is one of the factors that generates spiral arms in galaxies; it also plays a useful role in keeping the contents of the galaxy stirred up together.

4.3 Measuring the mass of a galaxy

There is more to the 21-cm line than beautiful images. By making the right sorts of measurements we can use 21-cm data to measure the masses of both the atomic gas and the stars in Messier 83. To understand how to do this we need to look more carefully at what causes an atom to emit a 21-cm photon.

The 21-cm photons detected by radio astronomers are the result of *spontaneous* downward transitions from the parallel to the antiparallel state. It is not possible to predict when a spontaneous transition will take place in a particular atom, but for many cases the statistical probability that a transition will take place within a given time can be measured or calculated using quantum mechanics. Some types of transition occur very quickly and are given the name **permitted transitions**; hydrogen atoms in the $n=2$ energy level, for example, spend only about 10^{-8} seconds before dropping to one of the two $n=1$ states. **Forbidden transitions**, on the other hand, have a much lower probability of occurring within a given time, and an atom may stay in its excited state for years before dropping to the lower level. The 21-cm transition is one of the most 'forbidden' transitions known; left to itself an atom in the parallel state will wait something like 10 million years before making its move and emitting a 21-cm photon. Another way of expressing this idea is that on average, one 21-cm photon per year is produced from every 10 million upper-state atoms in a cloud.

Although 10 million years is a long time in human terms it is short compared to the 16 billion year age of the Galaxy. Why, then have not the interstellar hydrogen atoms all dropped to the ground (antiparallel) state? The answer is that, even at the extremely low particle densities encountered in interstellar space, collisions between hydrogen atoms are frequent enough to shift atoms back into the upper state. Under typical interstellar conditions one hydrogen atom will collide with another roughly every thousand years. These collisions frequently trigger both upward and downward transitions between the two quantum states. Since a thousand years is much shorter than 10 million years almost all the transitions between the two states will be collisional. Under these circumstances it can be shown by the application of quantum mechanics that at any given time just over one quarter of the hydrogen atoms are in the lower state while just under three quarters are in the upper state.

What this somewhat circuitous argument tells us is that the number of 21-cm photons per second emitted by a gas cloud or by a galaxy depends only on the total number of hydrogen atoms in the cloud. So long as there is no absorption (see Section 4.6) the strength of the 21-cm line is practically independent of the density or the temperature of the hydrogen. If we know the distance to a galaxy we can calculate the total mass of atomic gas directly from the measured strength of the 21-cm line. For Messier 83 this mass turns out to be the equivalent of 10 billion stars the mass of the Sun – usually written as 10^{10} $M_\odot$. The simplicity of the relationship between line strength and mass is one of the reasons that the 21-cm line is so useful. Most of the other spectral lines we shall meet in this book are much more difficult to interpret.

Atomic hydrogen is only one constituent of a galaxy. How can we 'weigh' the rest of the matter in it, such as its stars? The answer is to measure the gravitational force that is holding the galaxy together. Each atom in a galaxy feels the combined gravitational force of all the other atoms in the galaxy. The net result is a force pulling it towards the center, where the galaxy is most concentrated. In a stable galaxy this inward gravitational force is balanced by the outward centrifugal force that is produced by the galaxy's rotation. By a simple application of Newton's law of gravitation (see Appendix I) we can estimate the total mass of a galaxy from measurements of its rotational speed at large distances from its center. In a spiral galaxy the gas and the stars revolve around the center at nearly the same speed. The Doppler shifts of the 21-cm line are therefore a good way for measuring the rotation. Applying this technique to M83 gives a *total* mass of 1.6×10^{11} $M_\odot$.

If we put these two mass estimates together we infer that the mass of atomic hydrogen is about 6% of the total mass in Messier 83. What is the nature of the rest of the mass that is not atomic hydrogen? A few percent is made up of either ionized or molecular hydrogen, neither of which emits the 21-cm emission line. Other kinds of interstellar matter, including helium and dust grains add up to about another 2%. The bulk of the remaining 90% or so is comprised of stars of one kind or another, although, as we will discuss in the final chapter of this book, there is a possibility that some of the mass may be in the form of 'dark matter' whose nature is still a mystery.

This ratio of 6% for the fraction of a galaxy's mass in the form of atomic hydrogen should be treated with some caution. The ratio varies from galaxy to galaxy and from place to place within a galaxy; generally the outer regions of a galaxy have a higher ratio of gas to stars than the inner regions.

4.4 Tools of the trade: radio astronomy

Making a map such as that in Figure 4.2 requires a far more sophisticated radio telescope than was available when the 21-cm line was discovered in 1951. At that time the science of radio astronomy was still in its infancy with the first detection of celestial radio signals of any kind having been made only in 1936. The discovery of the 21-cm line was made using a simple radio telescope consisting of a 20-m diameter metal reflecting surface which focused the radiation onto a radio receiver. With this simple arrangement radiation could be collected

Figure 4.3. The 100-m diameter radio telescope near Bonn, Germany. The large parabolic surface reflects radio waves to a radio receiver at the focus of the dish. A computer controls the telescope and allows it to point to any position in the sky.

from only one point in space at a time; to make a map of a galaxy or gas cloud it was necessary to aim the telescope at one point in the sky after another and put together a map after the observations were complete. Radio astronomers had nothing analogous to a photographic plate that could register a whole image at one time.

Scanning the sky with a single dish is not only laborious, it is also imprecise. The ability of an optical instrument to discern details in an image is referred to as its **resolving power**. The wave-like nature of electromagnetic radiation sets limits to the resolving power of a telescope and causes maps made with them to be blurred. The finest details visible subtend an angle of about $60(\lambda/D)$ degrees where λ is the wavelength of the waves and D is the telescope diameter. This so-called **diffraction limit** has a value of about half a degree for a 20-m dish at 21 cm; no detail finer than the diameter of the full Moon can be seen by such a system.

The quest for increased resolving power has been the major driving force in radio astronomy technology over the past four decades. One approach that has been used is to observe at higher frequencies, and radio waves as short as 0.3 mm wavelength can now be observed with specially-designed parabolic antennas. This approach does not help for observations at a fixed wavelength like 21 cm, however. Another way to get better resolving power is to increase the telescope diameter. The largest fully steerable antenna has a diameter of about 100 m, though larger ones have been built that have more limited maneuverability.

By far the most powerful technique for achieving high resolving power in radio astronomy is **aperture synthesis**, developed by astronomers in Cambridge, England in the 1960s. In this technique, signals are received simultaneously by a number of antennas separated by distances of several kilometers. As the rotating Earth carries the antennas around each other their signals are combined electronically to build a map of the sky with a resolving power equivalent to that of a dish several kilometers in diameter. Such a technique has been used to map details as fine as 0.1", far smaller than can be seen with the naked eye, and better than can be achieved with an optical telescope. Large aperture synthesis telescopes have been built in the United Kingdom, in the Netherlands, in the USA and in Australia. Aperture synthesis requires the use of very powerful computer systems to transform the received electrical signals into maps of the sky. Even higher resolutions have been achieved by combining signals from telescopes scattered around the world in a technique called **Very Long Baseline Interferometry** (VLBI). The VLBI technique was used to make the map of the cosmic masers in Figure 8.12.

Figure 4.4. The most powerful radio telescope in the world is the Very Large Array (VLA), comprising 27 parabolic antennas spread over the New Mexico desert. Each 25-m diameter antenna can be positioned along one of three 20-km long railroad tracks extending in a 'Y' pattern across the desert floor.

Aperture synthesis is now used for making most radio astronomy observations at wavelengths of 1 cm or more. At wavelengths shorter than this there are still technical difficulties in combining the signals from separate antennas; much millimeter-wave astronomy is therefore still done with single parabolic dishes. We will explore some of the technical problems of astronomy at wavelengths of around a millimeter in Chapter 8.

4.5 Atomic hydrogen in the Milky Way

The atomic hydrogen in our Galaxy is much easier to detect than the gas in Messier 83, but the observations are much harder to interpret. We have what amounts to a bird's eye view of Messier 83, but are like lost hikers in a forest when trying to study the Milky Way Galaxy. We can detect atomic hydrogen all around the Milky Way and we can measure how fast it is

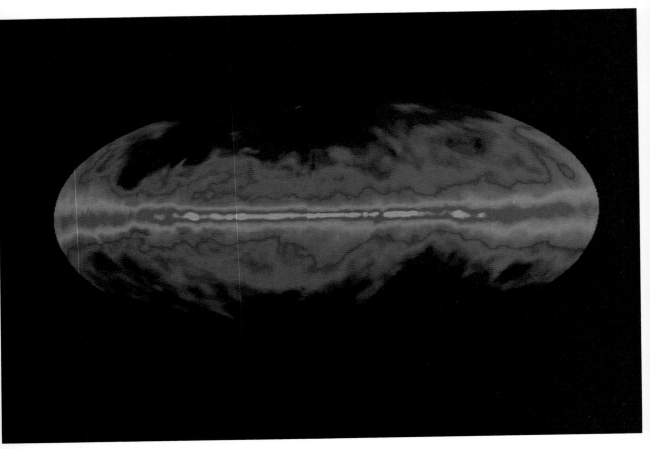

Figure 4.5. Emission from the 21-cm line is strongest along the plane of the Milky Way (red and white) than elsewhere in the sky (blue and black). From this we conclude that the galactic plane has the form of a thin sheet of gas. Compare this picture with the visible and infrared appearance of the Milky Way seen in Figure 1.5.

moving, but we usually have only a vague idea of how far away it is. A further complication is that there are great gaps in our knowledge of the shape of our own Galaxy, and that much of what we do know about it comes from studying the 21-cm line itself. Consequently, discussion of the 21-cm line is inseparable from a discussion of the structure of the Galaxy.

Emission at 21 cm is heavily concentrated in a narrow strip along the Milky Way. More than a few degrees away from the Milky Way the emission is much weaker. From this observation we deduce that the atomic hydrogen in the Milky Way, like many of the stars, is spread out in a flat sheet around us. This sheet is referred to as the **galactic plane**. We get very little idea of the shape or size of this sheet just from looking at how bright it is in different directions. It is the pattern of Doppler shifts in the 21-cm line that gives us the information we need to make maps of the Galaxy. The reason why the 21-cm line so useful is that radio waves suffer far less from absorption than do light rays. Distant stars in our Galaxy are hidden from us by dust clouds, but it is easy to measure the Doppler shifts of 21-cm photons from the farthest points in the Galaxy.

When we look at the 21-cm emission from different parts of the galactic plane two things stand out. First, the emission is strongest towards the direction of the constellation Sagittarius. Second, on one side of this peak the gas is predominantly redshifted, while on the other it is predominantly blueshifted. These data provide strong evidence that we live towards one edge of a rotating disk galaxy whose center is in the Sagittarius direction. The best current estimates are that the distance to the center is 8500 parsecs,

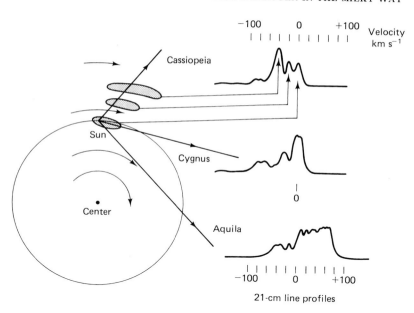

Figure 4.6. *The complicated profiles of the 21-cm emission lines from the galactic plane are the result of Doppler shifts caused by the differential rotation of the Galaxy. Velocities of redshifted objects are conventionally positive; those of blueshifted objects are negative. Hydrogen at different distances gives rise to emission features at different velocites. In some directions, such as Cassiopeia, the emission from different spiral arms may be separated quite easily, but in directions like Cygnus and Aquila the profiles are much harder to interpret.*

and that the Sun is moving in an almost circular orbit at a speed of 220 km s^{-1}. At this rate it makes one revolution around the Galaxy in 240 million years – about the time that has elapsed since the age of dinosaurs on Earth.

Looking for spiral arms in this rotating disk is much harder for the Milky Way than it was for Messier 83. The trouble is that there is no *direct* way of measuring the distance to a patch of atomic gas in the Galaxy. An astronomer has to proceed by looking for patterns in the 21-cm data that could be attributed to a spiral arm, making various guesses as to the shape and distance of this arm, and then seeing which guess fits the data best. This process is usually referred to as **modeling**, although the models usually exist only as mental images, equations, or computer programs.

Figure 4.6 shows the appearance of the 21-cm line emission line in different directions in the galactic plane. Spectral lines with profiles as complicated as these are usually referred to as **spectral features** rather than lines. When looking towards the constellation Cassiopeia we find that the hydrogen appears to be moving towards us with speeds ranging from zero to -80 km s^{-1}. These blueshifts occur because of differential rotation. The outer parts of the Galaxy orbit at a slower rate than the Sun; we are therefore 'overtaking' the gas that is farther out from the center. In the Cassiopeia direction the blueshift of the atomic gas increases with the distance from the Sun; the shape of the 21-cm line therefore indicates approximately how the gas is spread along the line of sight. The peak at 0 km s^{-1} is due to gas in our vicinity, moving at the same velocity as the Sun, while those at -20 and -40 km s^{-1} are due to concentrations of gas in spiral arms at distances of approximately 2000 and 4000 parsecs respectively.

The 21-cm emission line shapes in other directions are generally harder to interpret than those towards Cassiopeia. One major problem is that in some directions, such as towards the constellation Aquila, it is possible for clouds at different distances to share the same redshift. The most detailed impression we have of the spiral pattern in our Galaxy is shown in Figure 4.7. It indicates that the Galaxy contains a number of thin spiral arms, wound tightly around the center, but because of the difficulties of determining the distances to atomic hydrogen clouds, details of the map should not be taken too seriously. Although the Milky Way Galaxy is certainly a

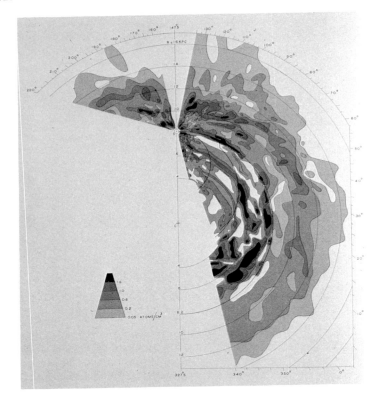

Figure 4.7. *The spiral arms in the Milky Way as deduced from observations of the 21-cm emission line. A certain amount of intelligent guesswork was involved in the preparation of this figure. Only the parts of the Milky Way visible from the northern hemisphere are displayed in this picture.*

spiral, its arms appear to be less prominent than those in some other galaxies such as NGC 2997 (Figure 1.1) or Messier 83 (Figure 4.2).

We can use 21-cm spectra to investigate the thickness of the atomic gas layer in our Galaxy, as well as its spiral shape. Like the atmosphere of the Earth, the interstellar gas is pulled into a thin layer by the force of gravity. In our own atmosphere the gravity is the pull of the Earth itself; in the Milky Way the force comes from the billions of stars in the disk which act to pull the gas towards the middle of the galactic plane. It is a simple exercise in physics (see Appendix J) to calculate how a gas layer behaves under the force of gravity. What one finds is that the density of such a layer steadily decreases with increasing altitude without any sharp edge ever being reached. As one moves up through the layers of such an atmosphere the net weight of the gas remaining above one's head continuously decreases. With less downward pressure on it the gas is able to expand, causing its density to drop. The resultant atmosphere has a characteristic thickness which is called the **half-height**. For every half-height one rises in the atmosphere the density drops by a factor two. The half-height of the Earth's atmosphere is 6 km. The half-height of the atomic hydrogen in the galactic plane is 180 parsecs. Since the diameter of the Galaxy is about 30 000 parsecs, and the gas layer is only a few hundred parsecs thick, its proportions are something like those of a long-playing record.

Once we know how far away a patch of atomic hydrogen is, we can calculate its mass and its average particle density using the same method as we employed for the galaxy Messier 83. We find that the total mass of the atomic hydrogen is about five billion times the mass of the Sun. Averaged over the whole galactic disk this mass corresponds to approximately one neutral hydrogen atom per cubic centimeter (1 cm^{-3}) or one solar mass every fifty cubic parsecs.

4.6 Interstellar clouds

Knowledge of the average particle density tells us little about how the gas is spread out in space. From the 21-cm emission lines alone we cannot tell whether the atomic hydrogen exists as a collection of relatively dense clouds or as a smooth sea of gas. Neither can we tell what the temperature of the emitting gas is. Both these questions can be addressed by studying the way that 21-cm radiation from a distant radio source is *absorbed* as it passes through atomic hydrogen.

As explained in Chapter 3, radiation passing through a region of space containing hydrogen can cause *stimulated* transitions between the two lowest energy states. If the transition is from the lower to the upper state a 21-cm photon is absorbed; if it is from the upper to the lower an extra 21-cm photon is emitted. From quantum mechanical calculations it can be shown that a 21-cm photon is exactly three times more likely to stimulate an upward transition than a downward transition. This factor three is almost exactly counterbalanced by the greater number of atoms that are pushed into the upper state as a result of collisions between atoms. The balance is not perfect, however, and there are slightly more stimulated absorptions than emissions. The imbalance becomes greater at lower temperature, as the fraction of atoms in the upper state drops farther from its maximum possible value of 75%. In other words, cold gas absorbs more 21-cm radiation than warm gas; quantitatively it can be shown that the amount of 21-cm radiation absorbed by atomic hydrogen is inversely proportional to its temperature.

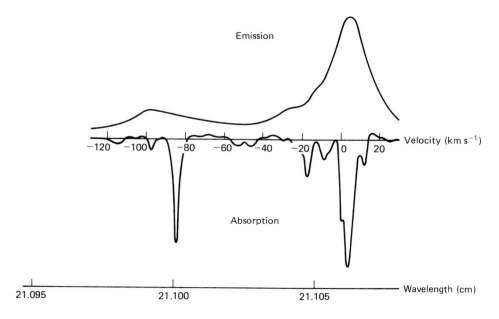

Figure 4.8. The 21-cm absorption line produced by atomic hydrogen in front of the distant radio galaxy Cygnus A. Cold gas absorbs much more strongly than warm gas; the strongest absorptions, such as those at +4 and −85 km s^{-1}, occur in interstellar clouds where the gas temperature is about 80 K. A 21-cm emission line in almost the same direction shows fewer peaks and valleys than the absorption line because both warm and cool hydrogen contribute to it equally.

To study absorption of the 21-cm line we need to point a radio telescope through the galactic plane towards a source of continuum radio emission that is well outside the Galaxy. The distant radio galaxy Cygnus A acts as a suitable beacon. Figure 4.8 shows the spectrum of the radio emission from Cygnus A near 21 cm, showing a series of absorption lines with different Doppler shifts. Also shown is a 21-cm emission spectrum measured along a line which just skirts Cygnus A. The differences between the two profiles are striking. Whereas the absorption spectrum shows a series of narrow features, notably at velocities $+4$, -11 and -85 km s^{-1}, the emission spectrum is comparatively smooth over the whole velocity range $+20$ to -120 km s^{-1}. The existence of these narrow features implies that at certain positions along the line of sight to Cygnus A are concentrations of gas with much lower than normal temperatures. These concentrations are usually referred to as **cool clouds**. Their gas temperatures are typically about 80 K and their particle densities are in the range 10–100 atoms cm^{-3}. The gas which is seen in the emission line profile but which does *not* produce strong absorption features is known as the **warm neutral medium**. Its temperature is about 8000 K, but its density is only about 0.1 atoms cm^{-3}. The total mass of the cool clouds is roughly the same as the total mass of the warm neutral medium, but the clouds occupy a much smaller fraction of the volume of the Galaxy. Warm and cool atomic hydrogen provides another example of interstellar phases. The reason why astronomers use the rather weak words 'warm' and 'cool' to describe these phases of the interstellar medium is that there are other phases we will meet later in the book that are much colder and much hotter than those described in this chapter. In human terms, though, the words are gross understatements. The 'cool' clouds have temperatures nearly 200° below zero on the Celsius scale; the 'warm' clouds are hotter than the surface of the Sun.

It is very difficult to determine the shapes of the cool clouds in the Galaxy. The problem is that we can only distinguish between warm gas and cool gas in those comparatively few directions where there happens to be a suitable background radio source. These radio sources are so few and far between that we cannot tell where one cool cloud starts and another ends. Almost everything we know about the shapes of clouds has therefore come from studying 21-cm emission profiles, which can be measured in every direction in the sky. The disadvantage of using emission line profiles is that we do not know whether the gas we see is warm or cool.

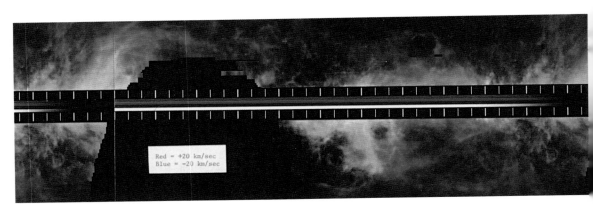

Figure 4.9. When we look away from the galactic plane the atomic gas has a wispy appearance that implies a filament-like or sheet-like structure for the interstellar medium. In this image red represents gas moving away from us, and blue represents gas moving towards us. The blank strip in the middle of the diagram shows the location of the galactic plane; the blank area to the lower left is the region of sky invisible from the northern California location of the radio telescope where these data were obtained.

It was once supposed that clouds were more or less spherical, cool concentrations of a few parsecs in diameter drifting in a warm sea of low density gas. We now believe that the true picture is more complicated than this. The unevenness of the atomic hydrogen can be seen in Figure 4.9 which shows gas away from the plane of the Galaxy in more detail than Figure 4.5. By avoiding data from the galactic plane and excluding gas with large Doppler shifts we can be fairly sure that all the emission shown in the figure is produced by gas within a few hundred parsecs of the Sun. We cannot distinguish cool from warm gas in this image, but it is nevertheless clear that much of the neutral gas exists in the form of loops, sheets and filaments rather than as isolated spheres. How these strangely shaped clouds form and evolve will be discussed in Chapter 11.

4.7 Gas motions in the interstellar medium

So far, the only gas motions we have considered are those that arise from the rotation of the Galaxy. During the course of this book we shall meet many other types of motion. Some are easy to detect, while others can be inferred only indirectly. Generally, any interstellar matter that emits or absorbs sharp spectral lines can have its Doppler shift measured. Most types of interstellar gas come into this category, but not cosmic rays or dust grains. Even when it can be measured, the Doppler shift gives us only incomplete information. It only measures the **radial velocity**, which is the component of velocity in the direction *towards* us or *away* from us. It tells us nothing about tangential or 'sideways' velocities the gas may have in addition to its radial motion. The only way we can measure tangential velocities is by comparing images of the sky taken a number of years apart; very little useful data on interstellar motions has ever been obtained by this method.

One of the most important motions we have to consider is the random thermal motion of individual gas atoms. There is a simple relation (see Appendix G) between the temperature of a gas and the average velocity of its atoms. At 80 K hydrogen atoms move at about 1 km s^{-1}; at 8000 K they move 10 times faster. These velocities are averages of course, with some atoms moving faster and some slower. Whenever we look at a spectrum line we would therefore expect it to be broadened by the Doppler shifts corresponding to these motions, at the very least. In practice we find that the 21-cm absorption lines are considerably broader than can be explained

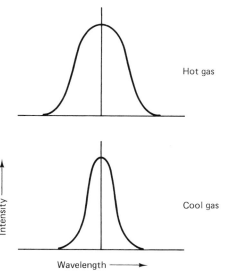

Figure 4.10. The random thermal motions of the atoms in a warm gas produce Doppler shifts in emission and absorption lines. These shifts give the spectrum line a characteristic bell-like shape.

by thermal motions alone. The gas in the clouds must therefore be moving in other ways as well. There are many possible broadening mechanisms which we will meet later in this book. The cloud may be rotating, or expanding or contracting. The cloud may consist of several cloudlets moving at different velocities, or may have somehow been churned up into turbulent motion.

Another important kind of gas motion we must consider is sound waves. A recent science fiction thriller was wittily promoted using the slogan 'In space, no-one can hear you scream'. It is true that sound waves travel very poorly in the thin gas of interstellar space, but they certainly do exist. The speed of sound in space depends mainly on the gas temperature, and is always about the same as the velocity of the thermal motions. In an 80 K gas of atomic hydrogen the sound speed is 1 km s^{-1}, approximately three times faster than the speed of sound in the Earth's atmosphere. It is important to realize that by astronomical standards this velocity is extremely slow. For comparison, the Earth's velocity around the Sun is 30 km s^{-1}, the Sun's velocity around the Galaxy is 220 km s^{-1}, and the velocity of gases ejected explosively from stars can exceed 10 000 km s^{-1}. A sound wave travelling away from us at 1 km s^{-1} since the creation of the Solar System would have barely reached the edge of Milky Way Galaxy by now.

Because the sound speed is so slow, most things move through space at **supersonic** velocities, meaning speeds faster than sound. Wherever supersonic velocities are found **shock waves** are produced. Shock waves are places where the velocity of a gas changes abruptly. On Earth a familiar example is the sonic boom that is heard when an aircraft flies faster than the speed of sound. In interstellar space they are commonly found where a stream of moving gas collides with a stationary cloud; the coherent motion of the gas stream is converted by the shock wave into random thermal motions. Energy of motion is thus converted into heat. Shock waves are therefore often associated with turbulence and with hot gas. In this book we will meet shock waves that are caused by the deaths of stars, by collisions between interstellar clouds and by the wind from the Sun. We will also see how shock waves can lead to the birth of stars, the rejuvenation of cosmic rays and the destruction of dust particles.

5 Ionized gas in the interstellar medium

As we learned in Chapter 2, atoms which have temporarily lost one or more of their electrons are described as ionized. The term is also applied to a gas if most of its atoms are in an ionized state; another word that is used to describe an ionized gas is **plasma**. The air we breathe consists almost entirely of neutral molecules and atoms. Except at altitudes above about 100 km, and in the immediate aftermath of a lightning strike, only a very tiny fraction of the Earth's atmosphere is ever ionized. Such ions as do exist are temporary, existing only for as long as they can avoid being neutralized by an oppositely-charged particle. On Earth, ionized gases are generally conspicuous only when they are artificially maintained; street lights and advertising signs are among the devices that depend on ionized gas for their operation.

Beyond the Earth's atmosphere, the story is quite different. Ionized gas, in some form or other, accounts for most of the mass, and fills most of the volume of the Universe. The Sun and the stars are almost fully ionized, as is the gas between the planets of the Solar System, and the gas between the galaxies. As far as we know, the only places in the Galaxy where matter is predominantly neutral are the planets, the comets, and parts of the interstellar medium.

At least half of the volume of our Galaxy is filled with ionized interstellar gas. In some locations the gas shines brightly, forming beautiful nebulae that are the most spectacular of all interstellar phenomena. Elsewhere, the ionized gas is so thin that it comprises the closest thing to a vacuum that exists in the Galaxy. To understand why ionized gas comes in this variety of forms, and how its existence is balanced against that of the atomic gas in the Galaxy we must take a look at the reasons why interstellar gas should be ionized at all.

5.1 Photoionization

To become ionized, a neutral hydrogen atom requires the input of at least 13.6 eV of energy (see Figure 3.3). It can get this energy in one of two ways. It can be **collisionally ionized** by being struck by another particle which is moving with at least 13.6 eV of kinetic energy, or it can be **photoionized** by absorbing a photon which has at least 13.6 eV of energy. Because of the connection between wavelength and energy such a photon must have a wavelength of less than 912 Å, in the ultraviolet range. Most of this chapter is concerned with photoionized gas, which is the type of gas that makes up most prominent nebulae. Very hot, collisionally ionized gas will be discussed at the end of the chapter.

Any photon with wavelength less than 912 Å is capable of ionizing a hydrogen atom, but the ionizing efficiency decreases as the wavelength of the photon gets shorter. Ionizations occur most readily at photon wavelengths between 912 Å and about 100 Å. Photons in this wavelength range are referred to as **ionizing photons**. The most important sources of ionizing photons are stars with surface temperatures more than 30 000 K. Such stars are comparatively rare; the surface of the Sun – a fairly normal star – is only at about 5800 K and produces very little radiation in this wavelength range. Stars whose ultraviolet radiation is powerful enough to ionize the surrounding interstellar medium are referred to as **exciting stars**. The word 'exciting' refers to the fact that much of the light from a nebula comes from atoms and ions in a nebula that have been excited into upper energy levels.

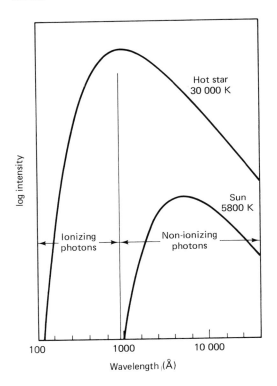

Figure 5.1. Continuous radiation from a 30 000 K hot star and from the Sun, whose temperature is near 5800 K. The Sun emits very little power at wavelengths shortward of 912 Å – the longest wavelength that can ionize hydrogen. Note that both axes in this figure are logarithmic; hence the difference in shape between these curves and those in Figure 3.8.

An ionizing photon travelling through neutral hydrogen is very quickly absorbed; the average journey of a 900 Å photon through a region containing 1 hydrogen atom cm^{-3} is only 0.05 parsecs – much less than the typical separation between adjacent stars. One important consequence of this strong ultraviolet absorption is that the interstellar medium around the Sun is essentially opaque in the 100–912 Å range, making astronomy at these wavelengths very difficult.

Although a 900 Å photon cannot travel far through neutral gas, it can pass almost unimpeded through a region of space that is already ionized. It can do this because individual protons and electrons interact only very slightly with ultraviolet radiation. A star that produces sufficient ultraviolet photons can therefore surround itself with a region of gas that is almost completely ionized. Within this region there is a balance between the rate at which protons and electrons recombine to form neutral atoms and the rate at which the neutral atoms become re-ionized by the ultraviolet photons from the exciting star. In a nebula with particle density 10^3 ions cm^{-3} a hydrogen atom would remain ionized for some 100 years before it meets an electron and recombines. It would then remain in a neutral state until struck by an ionizing photon; near to an exciting star the waiting time would only be a few months.

Because ionizing photons travel freely through ionized gas but are impeded by neutral gas, most of the interstellar medium is either almost fully neutral or almost fully ionized. This is why it makes sense to refer to different regions of space as H^0 regions or H^+ regions. The boundary between a region of ionized and a region of neutral gas is often comparatively sharp and is called an **ionization front**. The size of an H^+ region around a star depends on the density of the gas and on the amount of ultraviolet power being put out by the star. Around some stars the region of ionization extends far into the diffuse interstellar medium, but in other cases the ionized gas has the form of a comparatively dense nebula that surrounds the star and is physically associated with it. Where do these nebulae come from?

5.2 Exciting stars

Methodical studies of stars over the last 50 years have given astronomers a fairly clear picture of how stars work and how they evolve with time. Most of the stars in the Galaxy, including the Sun, are **main-sequence** stars; the energy that they radiate into space comes from the conversion of hydrogen into helium via thermonuclear reactions. The most important property of a main-sequence star is its mass. The more mass a star has the greater the pressure at its center and the faster the nuclear reactions will take place. This increase in nuclear power generation causes the star to become larger and hotter as well as more luminous. A star with 40 times the mass of the Sun (40 $M_\odot$) has 8 times its surface temperature, 14 times its diameter and 250 000 times its power output. High-mass stars such as these, which are often called **OB stars**, are the most powerful sources of ionizing photons in the Galaxy.

Main-sequence OB stars burn their hydrogen fuel so quickly that they

Figure 5.2. The Eagle Nebula, also known as Messier 16, is a classic example of a nebula that is kept ionized by recently formed OB stars. The pink color comes from the $H\alpha$ emission line of hydrogen. Some small dust clouds may be seen as dark patches in front of the visible nebula.

have short lifetimes. Whereas the Sun has enough hydrogen fuel to keep shining for a total of 10 billion years, a 40 $M_\odot$ star will exhaust itself in a few million years. Because of their comparatively brief lives, OB stars are rarely able to stray far from their place of birth. Consequently they tend to be found close to the remnants of the dense interstellar clouds in which they were born. When these clouds absorb the ultraviolet photons they become ionized forming a bright nebula that appears to surround the OB stars. The Carina Nebula (Figure 1.3), the Eagle Nebula (Figure 5.2) and the Orion Nebula (Figure 5.3) are classic examples of such a phenomenon. Since stars are always formed in groups it is common to find nebulae that are simultaneously being ionized by several OB stars. The phrase **H^+ region** is usually reserved for nebulae associated with recently formed stars.

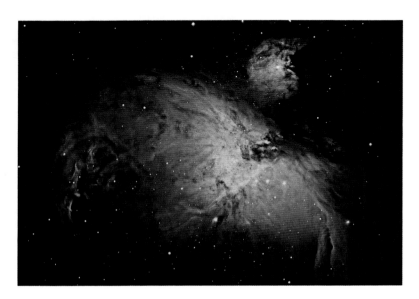

Figure 5.3. The Orion Nebula is the nearest bright H^+ region to the Earth, 500 parsecs away. It can be seen by the naked eye as a fuzzy patch of light in the Sword region of the Orion constellation. The gas is ionized by a small group of OB stars known as the Trapezium Cluster, which lies close to the brightest part of the nebula. Immediately behind the nebula, but invisible to optical astronomers, is a cloud of molecular hydrogen in which new stars are still being formed. The H^+ region is slowly eating its way into the molecular cloud as photons from the OB stars penetrate the molecular cloud, first dissociating the H_2 molecules and then ionizing the resultant neutral atoms. The molecular gas clouds are illustrated in Figure 8.7.

The other stars that are hot enough to radiate substantial amounts of ionizing radiation are the **white dwarfs**. As their name implies, these are physically much smaller than main-sequence OB stars. The word 'white' is a reference to their temperature – 'white-hot' stars being warmer than 'red hot'. White dwarf stars are extraordinary objects. They contain about as much mass as the Sun compressed into a volume about the size of the Earth. They represent the final stage in the evolution of a star, when it has exhausted all its nuclear fuel and can no longer maintain enough gas pressure in its interior to withstand the force of its own gravity. As will be discussed in Chapter 10, the formation of a white dwarf from a larger star is often accompanied by the ejection of material into space. When this material is ionized by the ultraviolet radiation from the white dwarf star a **planetary nebula** is formed. Planetary nebulae (which, despite their name, have nothing whatever to do with planets) are fairly easy to distinguish from H^+ regions by their greater symmetry and their isolation; they are manifestations of the lonely deaths of stars rather than their gregarious childhoods.

Figure 5.4. A planetary nebula is a cloud of ionized gas surrounding a white dwarf star. Planetary nebulae come in a variety of shapes; this one is given the name of the Helix Nebula. The gas has been expelled from the star at the center of the nebula and is now being lit up by the ultraviolet radiation from the star.

5.3 Tools of the trade: astronomical photography

Photography has been an integral part of astronomical research for over a hundred years, for recording both images and spectra. Photography has two immediate advantages over the human eye; it produces a permanent record, and it can record the light from much fainter stars than can be seen by the naked eye. Astronomical photography differs from regular photography in several ways. First, images are usually recorded on glass photographic plates rather than on film; being rigid, plates are stronger and suffer much less from distortion. Second, because celestial objects are faint, exposure times are usually measured in minutes or hours, rather that in hundredths of a second. Third, astronomers rarely use color film, since it is grainy and slow and reproduces the colors of H^+ regions poorly. When they want to study the color of an object they take several monochrome images with different color filters in front of the plate. Colors are then reconstructed from the different wavelength images, or simply calculated from the intensities measured on the different plates. Most of the color images in this book were made by combining images made at two or more different wavelengths.

The heyday of astronomical photography is over. For most purposes electronic cameras such as **charge coupled devices** (CCDs) are much more sensitive and convenient. In these instruments the photons are collected in an array of detectors on the face of a silicon chip. The image is immediately ready for study and storage in a computerized form; no messy chemical development is required. Electronic detectors have other advantages; they are sensitive at ultraviolet and infrared as well as visible wavelengths, they collect photons more efficiently, and they can provide a more accurate measure of the number of photons collected. Astronomers who study individual stars and galaxies are almost always better off with CCDs than with photographic plates.

The one application for which photographic plates are still better than CCDs is wide-field imaging. The largest CCDs used for astronomy contain about four million individual detectors in a 2000×2000 array. For a good quality image each detector should cover an angle of about 0.3" on the sky. The largest patch of sky that a CCD can image in one shot is therefore about 10' square, so any astronomer that needs pictures of

Figure 5.5. The Scorpius region of the Milky Way, showing numerous star clusters, dark clouds and bright nebulae. Photography is still the only way of obtaining images of large areas of the sky, although electronic cameras such as CCDs are often used for studying individual stars, galaxies and nebulae.

larger areas of sky than this still needs to consider photography. Research programs that need wide-field photographic images include surveying large areas of the sky for unusual objects, tracking the changing appearance of a nearby comet and tracing the faint wisps of large galactic nebulae. Most of the illustrations of nebulae in this book are derived from photographs rather than from electronic cameras. Photographic plates are in regular use that photograph 6° × 6° patches of sky. They are used in **Schmidt cameras** which are special telescopes designed exclusively for wide-field work.

5.4 The light from an H^+ region

For reasons we will discuss in the next section, the gas temperature in an H^+ region is always roughly 10 000 K. The light it emits is therefore in the form of emission lines rather than a continuous spectrum. The most important of these emission lines are those of hydrogen, helium and oxygen. The hydrogen and helium emission lines are sometimes called **recombination lines**, since they occur as a direct result of the recombination of an electron with a hydrogen or helium ion. When a recombination takes place the resultant neutral atom is usually left in an excited state rather than the ground level; it then cascades down to lower and lower energy levels, emitting a photon at each transition, until it reaches the ground energy level where it remains until it is ionized again by an ultraviolet photon. Hydrogen recombination lines are produced with a wide range of wavelengths, from the radio to the ultraviolet. The hydrogen lines that are seen at visible wavelengths form the so-called **Balmer series**, and arise as electrons in the upper states of a hydrogen atom drop to the second lowest

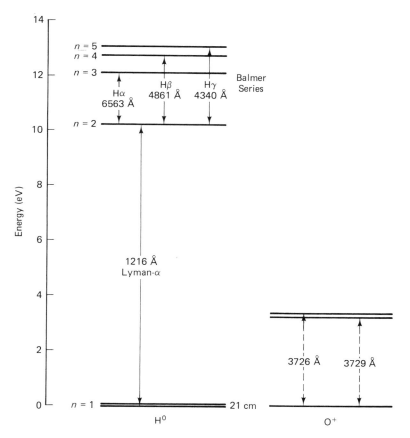

Figure 5.6. Energy levels of H and O^+. Only the lowest energy levels above the ground state are shown, and some details such as the splitting of the hydrogen $n \geq 2$ levels have been omitted. The O^+ ion has several energy levels that are closer to the ground state than the $n = 2$ level of hydrogen. Consequently it is easier for O^+ to become excited than for H^0. The 3726 and 3729 Å transitions in O^+ are 'forbidden', but they nevertheless give rise to two of the most intense emission lines from H^+ regions.

($n = 2$) state. The transitions in this series are labelled Hα, Hβ, Hγ, etc. (Figure 5.6). The strongest hydrogen recombination line of all is the so called **Lyman**-α line at 1216 Å. This ultraviolet transition is between the $n = 2$ and $n = 1$ energy levels

Helium is the second most common element, comprising about 10% of the atoms in the Universe. The ionization potential of helium is much larger than that of hydrogen (24.6 eV as opposed to 13.6 eV) so that only stars which produce a lot of photons shortward of 504 Å can ionize helium in the nebula. Consequently helium recombination lines are seen only in those nebulae which are associated with very hot stars (> 50 000 K). The relative strengths of the helium and hydrogen recombination lines in a nebula can provide an astronomer with a useful estimate of the surface temperature of its exciting star.

In many nebulae the strongest emission lines are not those of hydrogen or helium, but those of O^+ at 3726 and 3729 Å and O^{++} at 4959 and 5007 Å (see Figure 3.4). The prominence of these lines is remarkable not only because oxygen ions are a thousand times rarer than hydrogen ions but also because these particular spectral lines are extremely difficult to produce in a laboratory. When they were first observed in nebulae, these lines could not be identified with any known chemical element. For a while there was speculation that the lines were due to a hypothetical chemical element called 'nebulium', and it was not until 1927 that the lines were correctly identified as 'forbidden' transitions of oxygen ions. Since then, forbidden lines have been seen from other elements such as nitrogen, neon and sulfur. Up to a quarter of the power emitted by the OB stars in a nebula can wind up in these 'forbidden' ionic lines. They are strong as a result of the key role that they play in regulating the temperature of an H^+ region.

5.5 Forbidden lines from ions

The temperature of a gas, whether it is ionized or neutral, is set by a balance between the heat supplied and heat lost. If a gas absorbs energy faster than it can lose it, its temperature will rise. If it radiates energy away too fast, it will cool.

The gas in an H^+ region is kept hot by the energy from the exciting star, but the heating process is indirect. When an ultraviolet photon from the exciting star meets a neutral atom, exactly 13.6 eV of its energy is used for ionization. Any energy that is left over is transformed into kinetic energy which is shared between the ion and the electron. Because it is lighter than the ion, the electron gets the lion's share of this energy and accelerates off at high speed into the surrounding gas. As it collides with other electrons and ions in the gas it is slowed down, and its kinetic energy is shared among other ions and electrons, becoming part of the **thermal energy** of the gas.

For its temperature to remain stable the ionized gas must find a way of getting rid of the excess energy that the exciting star is depositing in it. The most efficient way of moving energy over large distances in space is by using photons, and the most efficient way of turning kinetic energy into photons is for fast moving particles to excite an atom or ion which then spontaneously reverts to its ground state. The question then becomes one of identifying which transitions are most important for carrying energy away from the gas. The main constituents of the ionized gas, the protons and the electrons, are of no direct use, since by themselves they do not have any discrete energy levels to be excited. Neutral hydrogen atoms (of which there are bound to be a few within the ionized region) have other problems; the lowest energy level above the ground state is the one that generates 21-cm photons. These photons have such a long wavelength that they do not play a useful role in carrying energy out of the H^+ region. All the other excited energy levels of hydrogen require at least 10.2 eV to be excited (see Figure 5.6). Collisions among gas particles are not violent enough to release this much energy unless the gas temperature is at least 30 000 K. Helium atoms and helium ions suffer from the same problem, namely that large amounts of energy are required to excite even the lowest of their upper energy levels.

When all possibilities have been explored we find that it is the ions of oxygen (O^+ and O^{++}) and certain other relatively common elements that have the most useful transitions for cooling the gas. An electron needs to have only 3.4 eV of kinetic energy to push an O^+ ion into one of it first two excited states (Figure 5.6). The 3726 Å or 3729 Å photons that are produced by the spontaneous downward transition from these states carry energy out of the nebula. Quite small changes in the gas temperature make very large differences in the rate at which the photons are produced by the O^+ and other ions; the rate increases by a factor five for a temperature increase from 8000 to 12 000 K. For this reason, the temperature of an H^+ region tends to stabilize at around 10 000 K even under quite large variations in the ultraviolet output received from the exciting stars. The forbidden lines act as a kind of thermostat that keeps all H^+ regions at around the same temperature.

Forbidden lines are extremely useful to astronomers because of what they reveal about the conditions in an H^+ region. A technique often used is to measure the relative strengths of two or more spectrum lines from the nebula under study. Different pairs or groups of lines are selected for measuring different properties. The relative strengths of different forbidden lines of the same ion can tell astronomers how frequently collisions are taking place in the gas, and hence about its particle density. The ratio of the strengths of the two ultraviolet O^+ lines at 3729 Å and 3726 Å, for example, changes from 1.5 at low densities to 0.4 at densities of above 10^4 particles cm^{-3}. Other forbidden line ratios can be used for other purposes; the ratio of certain lines of the O^{++} ion is a very sensitive indicator of the gas temperature, while the ratio of the strengths of the O^+ and O^{++} lines tells astronomers something about the average energy of the photons ionizing the nebula. As will be described in Chapter 6, the strengths of the forbidden lines of different elements provide astronomers with one of their major tools for analyzing the composition of the gas in galaxies.

One final point about forbidden lines remains to be addressed. If they are so powerful in H^+ regions, why had they not previously been seen in a laboratory? To answer this question we must recall the original meaning of the phrase 'forbidden transition', which we introduced in Chapter 4. It refers to a transition in which the atom or ion lingers in its upper state for a long time before dropping back to its ground state. The O^+ ion stays, on average seven hours in its upper state before emitting a 3729-Å photon. This is a trillion times longer than a hydrogen atom usually stays in the $n = 2$ state. In a low density nebula this delay does not matter, because for every collisional excitation there will eventually be a spontaneous downward transition, even if there is a long wait before it happens. If the gas density is more than about 1000 ions cm^{-3}, however, an O^+ ion risks getting hit by another particle while still excited. Such a collision is likely to push the ion back into the ground state before it has emitted its 3729-Å photon. The process is called **collisional de-excitation**; the denser the gas, the harder it is to produce 3729-Å photons.

In order to view a spectrum of oxygen ions in a laboratory scientists use a tube filled with gas and excite it electrically. Even though the gas is at a relatively low pressure, the density in a laboratory experiment is always far higher than that in a nebula. In the tube, collisions occur so rapidly that the 3726- and 3729-Å lines are completely collisionally de-excited and too faint to see; the emission lines seen from a laboratory gas are all permitted transitions. The reason these permitted transitions are not seen in the spectrum of an H^+ region is that more energy is required to excite them than is available from collisions in a 10 000 K gas.

5.6 Radio and infrared emission from H^+ regions

Most of the radio emission from an H^+ region has a continuous spectrum. As an electron moves through the ionized gas it experiences ever-changing accelerations due to electrical forces between it and the charged ions. Every burst of acceleration generates electromagnetic radiation over a wide range of frequencies. This kind of radiation is termed **free–free emission** because the electron remains unattached to the ion both before and after the encounter. The net result of all these collisions is a continuous band of almost uniform emission at radio and infrared wavelengths.

Radio telescopes can provide pictures of H^+ regions which are completely unaffected by interstellar extinction. Many important H^+ regions, particularly those on the far side of the Galaxy, are so heavily obscured that the radio emission is the only source of detailed information we have. Radio maps are also important for looking at the youngest H^+ regions – those surrounding very recently-formed OB stars. These small, dense H^+ regions may be still hidden inside the dust-filled clouds in which they were born.

H^+ regions produce emission lines at infrared and radio wavelengths, as well as continuum emission. At infrared wavelengths the brightest emission

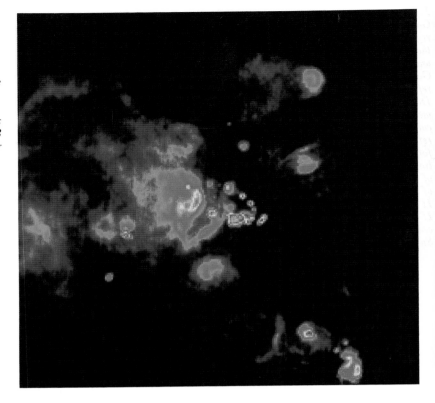

Figure 5.7. Radio emission from the W49 group of H^+ regions, which lies 14 kpc away in the constellation of Aquila. Because of extinction by dust in the Galaxy no trace of these objects can be seen at visible wavelengths. Each of the small H^+ regions contains at least one newly-formed OB star, and emits as much power as the Orion Nebula (Figure 5.3).

lines are certain forbidden lines such as the 12.8-μm line of ionized neon. At radio wavelengths it is the hydrogen and helium recombination lines that are the most prominent; the transition from the 110th to 109th energy level of hydrogen, which occurs at a radio wavelength of 6 cm, is a line that has been much used for studying obscured H^+ regions.

Often, the most useful piece of information that we get from a radio recombination line is its Doppler shift. The Doppler shifts of H^+ regions arise mainly as a result of differential rotation of the Galaxy. We can therefore estimate distances using the same sorts of ideas that we used in Chapter 4 to map out the neutral hydrogen in the Galaxy. For many obscured H^+ regions this method provides the only clue we have as to their distances. At optical wavelengths we see only the H^+ regions within about 2 kpc of the Sun, but at radio wavelengths we can see objects all the way to the other side of the Galaxy, 20 kpc away. When all the H^+ regions that have been measured are located on a map of the Galaxy we find that they tend to be concentrated in a ring at 4–6 kpc radius from the center of the Galaxy. Later, we will see that a similar distribution is found for the molecular gas in the Galaxy.

H^+ regions are among the most powerful sources of continuous infrared emission in the Galaxy. The emission comes not from the ionized gas but from intermingled dust grains which absorb the visible and ultraviolet light from the H^+ region and its stars. Infrared emission from dust is discussed in Chapter 7.

5.7 The warm ionized medium

Not all ionized gas resides in distinct nebulae. Very faint emission lines of hydrogen and sulfur can be detected from most directions in the Galaxy

Figure 5.8. H^+ *regions in the Galaxy. The distance of H^+ regions can be estimated from their Doppler shifts and an understanding of how the Galaxy rotates. The cicles refer to H^+ regions whose Doppler shifts can be measured using visible emission lines. The squares refer to obscured H^+ regions for which radio recombination lines must be used. The size of the symbol is an indicator of the power emitted by the H^+ region. Note that most H^+ regions are closer to the center of the Galaxy than the Sun is.*

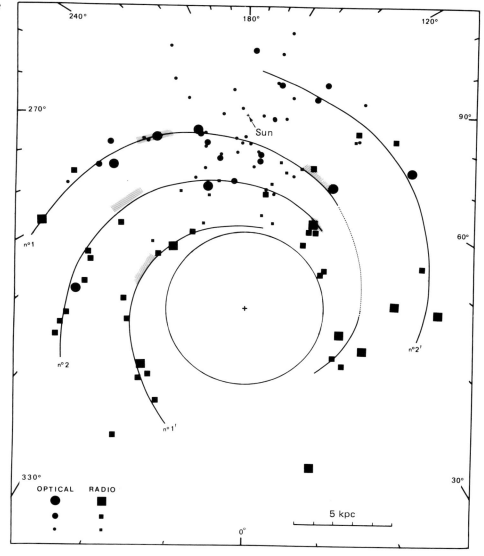

Special cameras with very narrow filters are needed to isolate these emission lines from other sources of light in the sky. The emission lines are sure signs that large volumes of ionized gas exist in the interstellar medium, even in locations that are not obviously associated with hot stars. The temperature of these ionized regions is estimated to be 8000 K, quite close to those of H^+ regions and to the warm neutral medium; this diffuse gas is usually referred to as the **warm ionized medium**; its temperature is similar to that of the warm neutral medium we discussed in Chapter 4. The average particle density of the warm ionized medium is of the order of 0.3 cm^{-3}, assuming that about 10% of the volume of the Galaxy is filled by it. The gas is probably ionized by OB stars that have moved well away from their birth sites, and by white dwarf stars which are no longer surrounded by recognizable planetary nebulae.

The warm ionized gas in the general interstellar medium reveals its presence in quite a different way. **Pulsars** are special kinds of stars in our

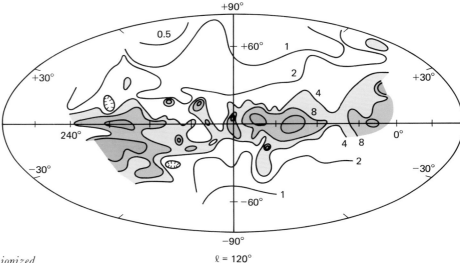

Figure 5.9. The warm ionized medium is revealed in this map of diffuse Hα emission around the sky. The plane of the Milky Way runs horizontally through the center of this picture. Extinction by dust stops the galactic plane from being as prominent in the Hα line as in the 21-cm line (Figure 4.5).

Galaxy which emit regular pulses of electromagnetic energy with periods ranging from a few milliseconds to a few seconds. By comparing the arrival times of pulses at different radio wavelengths it is found that radio signals are delayed slightly as they pass through the interstellar medium. The delay arises because electromagnetic waves travel slightly slower through space containing ionized gas than through a vacuum. The delay is greater for low-frequency radio signals than high-frequency ones; an 80 MHz signal from a 300 parsec distant pulsar is delayed by about 0.5 s by the interstellar medium, whereas a 150 MHz signal from the same pulsar is delayed by 0.15 s. This 0.35 s time difference is minute compared to the total time that a signal takes to traverse the Galaxy, but it can be accurately measured and used for studying the interstellar medium.

The ionized gas that causes the delays of pulsar signals is probably the same gas that produces the diffuse Hα emission lines, although there have been suggestions that some of the ions result from partial ionization of warm neutral gas.

5.8 Coronal gas

To compete our discussion of ionization in the interstellar medium we must briefly introduce the **coronal gas**. Coronal gas is an ultra-thin, ultra-hot gas that fills vast and otherwise empty volumes of interstellar space. Coronal gas is unlike the gas in H^+ regions in several fundamental ways. First, with a temperature of 1 000 000 K it is 100 times hotter and at least 100 times thinner than the photoionized gas we have been discussing in this chapter. Second, its ionization is caused by energetic collisions among the gas particles rather than by the absorption of ionizing photons from a star. Third, it is practically invisible as far as radio, infrared and optical telescopes are concerned; the only way of detecting it is by using specially-instrumented rockets or satellites.

Coronal gas is so elusive that we can safely ignore it for the time being. Its importance will emerge in Chapter 11, where we will try to explain how all the different phases of the interstellar medium interact with each other. We will find that coronal gas may turn out to be the most widespread form of interstellar matter in the Galaxy.

6 *The other elements*

Until now we have confined our attention almost entirely to hydrogen, the atom that dominates all discussions of the interstellar medium. It is now time to look at some of the other elements that are found in space. These elements will become particularly important when we get to discuss interstellar molecules and dust grains, but they can also play crucial roles as individual atoms. The mixture of elements in interstellar gas is known as its **chemical composition**. It is of interest to astronomers not only for its own sake but for what it reveals about the evolution of stars and about the history of the Galaxy.

The ratio of the number of atoms of an element to the number of atoms of hydrogen in the same volume of space is called its **relative abundance**. Astronomers determine the relative abundances of interstellar atoms by using the fact that each element has its own unique set of spectral lines. By comparing the strength of an element's spectrum lines with those of hydrogen one can, in principle, determine its relative abundance. Either emission lines or absorption lines can be used. As we shall see in this chapter, different techniques are best for different elements and for different regions of interstellar space. As a start, we will consider how to determine the abundance of helium (symbol He), the second simplest element.

6.1 Interstellar helium

The only way of determining the abundance of interstellar helium is by looking at the emission lines from H^+ regions. If a helium atom is ionized by a hot star it behaves in much the same way as hydrogen (Chapter 5). The He^+ ion moves freely about in the gas until it meets an electron to recombine with. When the recombination takes place the resultant He^0 atom is usually left in an excited state from which it cascades to the ground level, emitting a series of photons as it does so. As in the hydrogen atom, the resulting spectral lines are called recombination lines. So long as the central star of the H^+ region is hot enough to keep the helium almost all ionized, we can use the relative strengths of the helium and hydrogen recombination lines to calculate the relative abundance of helium.

It is found that in essentially all H^+ regions there is close to one atom of helium for every 10 of hydrogen. We have no reason to think that this ratio is any different in the diffuse interstellar medium. This result makes helium the second most abundant element in the interstellar medium (and, as we shall see, in the Universe as a whole). Since a helium atom weighs four times more than a hydrogen atom, the total mass of helium in the interstellar medium is about 40% of that of the hydrogen.

The high abundance of helium in space is all the more remarkable considering how rare it is on Earth. The element was first noticed in the spectrum of the Sun in 1868, but was unknown on Earth until 1895. It is a colorless, inert gas which does not form any molecules, even with itself. After hydrogen, it is the lightest gas that exists. Its low density and its disdain for other elements accounts for its rarity; since it refuses to combine with other elements it cannot form rocks or other solid substances, and since it is lighter than air it rapidly floats to the top of the Earth's atmosphere and escapes into space. The only places on Earth where helium is found in significant quantities are as a contaminant to the methane in underground natural gas fields. Despite its rarity, helium is of great commercial importance and for some purposes is virtually irreplaceable. Being the lightest non-flammable gas, it is ideal for inflating balloons. Being the most difficult substance of all to solidify, it is the best refrigerant for applications such as superconductivity which require very low temperatures.

The helium we harvest from underground gas wells comes from the radioactive decay of uranium in the Earth's rocks. As we will discuss in Chapter 10, interstellar helium has a quite different origin. While some was formed in the interiors of stars, most of it dates from the first few minutes of the Universe, and has remained unaltered since that time.

Helium is harder to ionize than hydrogen. Whereas the ionization potential of hydrogen is 13.6 eV, it takes 24.6 eV of energy to strip a helium atom of one of its electrons and a further 54.4 eV to remove the second one. Anywhere hydrogen is neutral – i.e. in H^0 or H_2 regions – we can be fairly sure that helium will be neutral also. Inside an H^+ region the situation is more complicated and depends on the surface temperature of the exciting star. Stars hotter than 40 000 K emit enough 24.6 eV photons to ionize essentially all the helium to He^+, but around cooler stars some of the helium has to remain neutral. Doubly ionized helium (He^{++}) is found only in certain planetary nebulae whose central stars can sometimes have surface temperatures exceeding 100 000 K.

6.2 Heavy elements in H^+ regions

There are a number of other elements whose abundance can be measured reliably in H^+ regions. These include oxygen, nitrogen, neon and sulfur. As we discussed in Chapter 5, ions of these elements can produce emission lines which may be as bright as those of hydrogen. The emission lines seen from these ions are not produced by recombination, but as a result of collisional excitations. Impacts by randomly-moving electrons in an H^+ region excite these ions into upper quantum states from which they make spontaneous downward transitions back to the ground level. The photons emitted as the ions drop to the ground state produce the strong emission lines seen from these ions. So long as the density of the ionized gas is not so high as to cause collisional de-excitation, we can calculate an ion's relative abundance from the ratio of the strength of its lines to those of hydrogen.

The best-studied H^+ region is the Orion Nebula (Figure 5.3). For every 1 000 000 hydrogen ions there is the mixture of atoms (summed over their various ions) shown in Table 6.1. What is remarkable about the data in Table 6.1 is the dominance of hydrogen and helium over all other elements. The third most common element, oxygen, is present only at the level of one atom for each 1600 hydrogen atoms. Taken together, hydrogen and helium comprise over 99% of the Universe; all the heavy elements, including such common materials as oxygen, carbon, silicon and iron, together account for less than 1% of all atoms.

Table 6.1. *Element abundances in H^+ regions*

Atomic Number	Element	Abundance
1	Hydrogen	1 000 000
2	Helium	100 000
6	Carbon	230
7	Nitrogen	40
8	Oxygen	620
10	Neon	70
16	Sulfur	30
17	Chlorine	0.1
18	Argon	1

From a humanistic point of view the four most common elements in H^+ regions after helium are a mixed bag. It is easy to accept that oxygen, carbon and nitrogen should be relatively abundant in space; oxygen and nitrogen make up most of the Earth's atmosphere while carbon is the basis of all living matter. What is galling is that neon, an inert gas whose main contribution to human culture is the advertising sign, should appear in such an exalted position in the table of elements. Like helium, neon forms no chemical compounds and is light enough that it easily escapes from the gravitational pull of the Earth.

There are many elements whose abundances cannot be determined from H^+ regions either because they have no suitable spectral lines at visible wavelengths or because the lines they do have are too faint. Silicon, magnesium and iron are among the relatively common elements that *cannot* be measured this way. Also, we cannot be sure that H^+ regions, which are necessarily associated with newly-formed stars, are representative of typical regions of interstellar space. What we need is a method of analyzing the gas in the *diffuse* interstellar medium. In most of space collisional excitations happen infrequently. Atoms therefore spend almost all of their lives in the ground energy level. The only transitions these atoms can make are *upward* transitions, in which a photon of the appropriate wavelength is absorbed, putting the atom temporarily into an excited state. What we have to do if we want to analyze diffuse interstellar gas, therefore, is to look for *absorption lines* corresponding to upward transitions out of the ground state. Unfortunately, the ground state absorption lines of most elements occur at ultraviolet wavelengths that do not penetrate the Earth's atmosphere. While a few elements, notably calcium and sodium, have been extensively studied by their visible-wavelength absorption lines, methodical studies of the composition of interstellar gas had to wait for the development of ultraviolet telescopes in orbit around the Earth.

6.3 Tools of the trade: ultraviolet astronomy

The human eye is blind to radiation shorter than 4000 Å, but the Earth's atmosphere is partially transparent to what is called near-ultraviolet radiation between 3000 and 4000 Å. Radiation in this waveband can be focused with ordinary ground-based telescopes and recorded on photographic plates. The 3726- and 3729-Å lines of O^+ (Figure 5.6) were discovered using photographic spectroscopy over 100 years ago, but serious studies of stars and interstellar matter at far-ultraviolet wavelengths – shortward of 3000 Å – did not start until the 1960s. The earliest far-ultraviolet experiments used small telescopes mounted in rockets. After being fired 100 km or so into space, these telescopes had about

10 minutes to make their pre-programmed observations above the atmosphere before falling back to Earth and (with luck) landing safely by parachute. Though the total amount of data sent back to Earth from these rockets was small, astronomers gained the experience they needed to design a series of sophisticated astronomical satellites that would open up the skies at far-ultraviolet wavelengths.

For astronomers interested in interstellar matter the most important of these ultraviolet observatories was the **Copernicus** satellite, which was launched in 1972 and which operated successfully until 1981. Copernicus contained an 80-cm diameter telescope which was operated by remote control from the ground as it orbited the Earth 15 times a day. Copernicus was designed for obtaining the spectra of stars between the wavelengths of 900 Å and 3000 Å. It worked slowly, staring at a single point in the sky for weeks at a time while its spectrograph progressively measured the brightness of the star at a long series of different wavelengths. The ultraviolet radiation was measured by electronic detectors that registered the arrival of individual photons into the telescope, and transmitted this information to the ground by radio. This painstaking approach resulted in extremely high quality spectra in which faint absorption lines and small Doppler shifts could be measured accurately. Almost all of the ultraviolet data alluded to in this book were obtained from this one satellite.

After the Copernicus satellite came the International Ultraviolet Explorer (IUE), launched in 1977 by a consortium from the USA, the United Kingdom, and the European Space Agency. Larger and more sensitive than Copernicus, IUE pushed the frontiers of ultraviolet astronomy well beyond the confines of the Milky Way Galaxy, giving astronomers their first clear sights of the ultraviolet spectra of quasars and other strange galaxies. IUE was a mixed blessing for students of interstellar matter, however. Its great sensitivity allowed astronomers to see interstellar absorption lines in the spectra of much more distant stars than was possible with Copernicus, but the spectra IUE obtained for bright stars lacked the great precision of Copernicus.

The banner of ultraviolet astronomy is now being carried by the Hubble Space Telescope (HST), launched in 1990. HST is a general-purpose unmanned astronomical telescope with a diameter of 2.4 m. It is fitted with five different instruments which can be changed from time to time by visiting astronauts. At launch the instruments included two electronic cameras and two spectrographs, all designed to operate both at visible and ultraviolet wavelengths. The initial performance of HST as a camera has been very disappointing, but as a spectrometer it has higher sensitivity and higher spectral resolution than either IUE or Copernicus, and will be extremely useful for studying absorption lines from interstellar gas. However it has the serious drawback that it cannot detect photons shorter than 1100 Å, and will therefore be unable to follow up on many of the discoveries made by Copernicus in the 900–1100 Å wavelength band.

Ultraviolet astronomy consists almost entirely of spectroscopy. Only a comparatively small effort has been put into obtaining ultraviolet images of celestial objects. One reason for this emphasis on spectroscopy is that ultraviolet radiation suffers much more from interstellar dust absorption than does light. Ultraviolet images are therefore usually harder to obtain and more difficult to interpret than visible images. The scientific rewards from ultraviolet spectroscopy greatly outweigh those from imagery.

6.4 Abundances from visible and ultraviolet absorption lines

Assuming that one has a suitable instrument at one's disposal, measuring a relative abundance is simple in principle. The telescope and its associated spectrograph are pointed at a star which lies beyond the region of space under study. Spectra are recorded at wavelengths where absorption lines are expected. By comparing the strengths of the absorption lines of the element with those of hydrogen an astronomer can calculate its relative abundance. In practice, there are a number of difficulties.

First, it is important to distinguish between absorption lines which are produced by interstellar gas and those which arise in the atmosphere of the background star. Fortunately, these two types of spectral line may usually be distinguished by their widths. Stellar absorption lines, which arise in the hot surface layers of stars, suffer more thermal broadening than those from the cool interstellar gas. In many stars the stellar lines are further broadened by the Doppler shifts caused by the star's rotation.

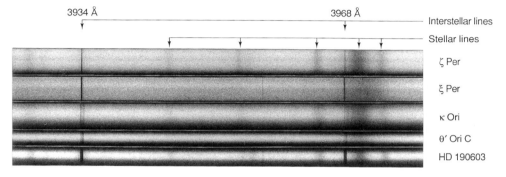

Figure 6.1. Interstellar calcium (Ca^+) absorption lines in five stars. Interstellar lines are much sharper than those produced in the atmospheres of stars. The double absorption lines in the spectrum of the star κ Ori are caused by interstellar clouds with different Doppler shifts.

Second, there is the problem of **saturation** of an absorption line. When there are relatively few atoms of a given type we can safely assume that the strength of its absorption lines are proportional to the total number of atoms doing the absorbing. If there are too many atoms trying to absorb at the same wavelength, however, there will eventually be no radiation left to absorb. Astronomers deal with the problem of saturation by trying to measure several different lines of an element, and, by concentrating on the faintest lines they can detect, where saturation is weakest.

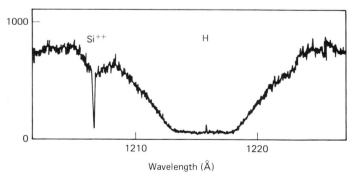

Figure 6.2. Saturated and unsaturated absorption lines. There are so many neutral hydrogen atoms in the direction of the star ζ Ophiuchi that the 1216-Å Lyman-α line is heavily saturated. Essentially all the light between the wavelengths of 1213 Å and 1219 Å is absorbed. When this happens the strength of the absorption becomes difficult to estimate. The narrow unsaturated line at 1206 Å arises from doubly ionized silicon.

Third, even in regions of space where the hydrogen is mostly atomic, certain elements are usually found in an ionized state, and an astronomer trying to estimate the relative abundance of an element must look for the spectral lines of its ions as well as of its neutral atoms. The elements which are most likely to be ionized are those with ionization potentials less than that of hydrogen, such as carbon, calcium and magnesium (see Appendix E). These atoms are ionized by ultraviolet photons in the 912–3000 Å wavelength range from hot stars spread throughout the Galaxy. Atoms with ionization potentials greater than that of hydrogen, such as nitrogen and helium, are usually neutral.

Fourth, only very limited regions of interstellar space can be analyzed by ultraviolet spectroscopy. For one thing, the only gases that can be studied are those which happen to lie along a line between us and an appropriate star. There are less than 100 stars that are bright enough for this kind of spectroscopy, so gas can be analyzed only in certain specific directions. A more serious problem is that most of the stars that have been studied are within a few hundred parsecs of the Sun. Stars farther away than this are generally too faint for ultraviolet telescopes to analyze in detail. Consequently ultraviolet spectroscopy tells us little about how the mixture of elements varies in different parts of the Galaxy, or from one galaxy to another.

Fifth, ultraviolet spectroscopy cannot be used to study relative abundances in *dense* interstellar clouds, because the dust grains that are always found in these clouds make the background star too faint for adequate data to be collected.

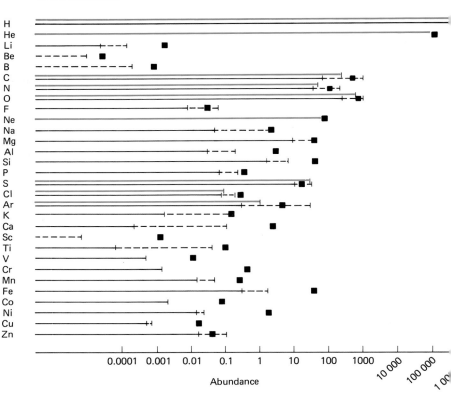

Figure 6.3. Interstellar and Solar System abundances. The lengths of the black lines indicate the numbers of atoms of each element in the interstellar medium for each 1 000 000 atoms of hydrogen. The dashed parts of the black lines indicate the ranges of values observed. The red lines show the abundances measured from H^+ region emission lines. The squares show the abundances measured within the Solar System.

Despite these problems, ultraviolet spectroscopy has given us interstellar abundances for nearly 30 elements. The results are shown in Figure 6.3. The elements are listed in order of their atomic number. The length of the black line is a measure of the abundance of that element as measured in the interstellar medium by ultraviolet astronomy, while the length of the red line gives the data for H$^+$ regions from Table 6.1. As in that table these abundances are given relative to 1 000 000 atoms of hydrogen, and include all ionization states. The dashed line at the end of the black line shows the range of values that have been determined for the interstellar abundance of that element. To some extent these ranges represent real variations in the chemical composition of the gas in different regions of space, and to some extent they represent uncertainties inherent in the measurements. Two elements, **He**lium and **Ne**on, do not have any suitable spectral lines for this kind of spectroscopy, and a few others (**Be**ryllium, **B**oron, **Sc**andium, **V**anadium) are too rare to measure.

Several important results stand out from Figure 6.3. First, we see that abundances in the diffuse interstellar medium are much the same as those in H$^+$ regions for those elements that have data in common. Hydrogen and helium are by far the most dominant elements, followed by a group consisting of **C**arbon, **N**itrogen, **O**xygen and **Ne**on. The next most common elements are **S**ulfur, Magnesium (**Mg**), **Ar**gon, **Si**licon and Iron (**Fe**). Argon is another inert gas like helium and neon, but the other four elements are solids at typical Earth temperatures. They readily react with each other and with other elements to make compounds such as rocks and minerals. The overall trend in the table is for complicated elements (i.e. high atomic number) to be rarer than simple ones, but there are some interesting exceptions; **Li**thium, **Be**ryllium and **B**oron are much rarer than many heavier elements, and even-numbered elements are generally more abundant than odd-numbered ones.

In trying to make sense of the abundances of interstellar gas we should note that there are two quite different questions we should be asking. First, why do the elements in the Universe come in the proportions that they do? We will postpone addressing the first question until Chapter 10, where we discuss the origin of interstellar matter. The other relevant question is, how and why do the abundances in the interstellar gas differ from abundances in other objects in the Universe?

6.5 Abundances in the Solar System

There are a few elements, such as oxygen and nitrogen, whose abundances have been measured in many different kinds of astronomical objects, including distant galaxies. The great bulk of our knowledge about the abundances of the elements, however, comes from studies of objects within the Solar System. The Solar System was formed out of an interstellar cloud some 4.6 billion years ago. The mixture of elements in the Solar System should therefore be a guide to the composition of the interstellar medium (gas plus dust) as it then was.

The easiest Solar System object to study is the Earth. Unfortunately, it is totally unreliable as a guide to the relative concentration of elements in the early Solar System. For one thing, its gravity is so weak that most of its hydrogen and helium atoms have long ago drifted away into space, leaving behind the rocky-metallic residue that forms the bulk of the planet. Secondly, the Earth's composition varies enormously from one place to another, as a result of nearly 5 billion years of repeated melting, recrystallization, sedimentation, and other geological processes. No terrestrial mineral can be said to typify the composition of the crust of the Earth, yet alone that of its core, mantle, ocean or atmosphere. For similar reasons, the rocks of the Moon are an unreliable guide to the composition of the material out of which it was made.

The two important sources of abundance data in the Solar System are the Sun, and certain meteorites. Solar abundances are determined by measuring absorption lines produced by the gases in the atmosphere of the Sun (part of the Sun's absorption line spectrum is shown in Figure 3.5). The theory is more complicated than that used for interstellar abundances because the gases have a range of temperatures, and emit light as well as absorb it. Many more absorption lines can be seen in the Sun's spectrum than in that of the interstellar medium, and enough have been identified that all but a handful of the 92 naturally-occurring elements have been detected and measured in the spectrum of the Sun. Strictly speaking, these resulting relative abundances refer only to the gases in the Sun's outer layers. Enough is known about the way that the Sun works, however, that we can be confident that, with a few interesting exceptions, the elements occur in similar ratios throughout the Sun's interior. We can also be fairly sure that no atoms in the Sun are hidden away in dust grains or the like because the Sun is hot enough everywhere to evaporate all known (and even theoretically conceived) solids and liquids.

Meteorites are lumps of interplanetary material that land on the Earth. Many tons of meteoritic material reach the Earth each day, but only a small fraction lands on the surface in pieces large enough to be recognized by a geologist. There are two properties of meteorites that make them very attractive to scientists studying element abundances. First, they can be taken into a laboratory and subjected to the full panoply of modern chemical analytical techniques. Secondly, analysis of the radioactivity in meteorites has shown that they all solidified close to 4.6 billion years ago, making them older than any known rocks on the Earth or Moon. They date from shortly after the Solar System formed out of an interstellar cloud, and they have orbited the Sun unchanged since then. Being small objects, they cooled quickly and retained no atmosphere; they have therefore been free of forces like erosion or volcanos that have distorted the historic record in the Earth's rocks. Several different kinds of meteorite can be distinguished. Some are mostly iron and others are mostly rock. The ones that are of most interest are the carbonaceous chondrites, since they appear to be the least changed since their formation.

Meteorites provide excellent abundance data for elements which normally exist in solid form, even for very rare elements like uranium which are a billion times rarer than oxygen. They are no good, however, for elements like neon or nitrogen which are gases at the temperatures of interplanetary space. Our best estimates for the overall abundances of the elements in the Solar System are therefore based on a combination of solar and meteoritic data. These values have been plotted in Figure 6.3 for comparison with the interstellar values. Elements 31–92 are omitted because they are so rare that very few have ever been detected in the interstellar medium; none have Solar System abundances larger than 0.01 on the scale of Figure 6.3.

The Solar System abundances plotted in Figure 6.3 are believed to be a fair representation of the mixture of elements in our Galaxy, at least as it was about 5 billion years ago. Astronomers use these values as a yardstick against which to compare other abundances – so much so that they are often rather grandly referred to as **cosmic abundances**. How do they compare with interstellar gas abundances? As can be seen in Figure 6.3, in the broad sense, the pattern is the same, with hydrogen and helium dominant, lithium, boron, and beryllium very rare, and a general decline in abundance with increasing atomic number.

When we look in more detail at Figure 6.3 we see some significant differences, notably that many elements are less abundant in the interstellar gas than in the Solar System. This phenomenon is called **interstellar depletion**. The differences can be dramatic; **Ca**lcium, **Ti**tanium, and iron

(**Fe**), are each 100 times rarer in interstellar gas than in the Solar System. Smaller, but still significant depletions are observed for elements like sodium (**Na**) and **Si**licon. Why are these particular elements depleted from the interstellar gas? An important clue comes from noticing that elements that are *not* depleted include the inert gases helium, neon and argon. Those that *are* heavily depleted, on other hand are usually found in solid form, at least at room temperatures. Could it be that the elements missing from the interstellar gas are in a solid form where they would be unable to produce ultraviolet absorption lines? To answer this question we have to turn our attention to a substance that we have up till now largely ignored, namely interstellar dust.

7 Interstellar dust

Dust grains have an importance in astronomy out of all proportion to their size. Although they comprise less than 1% (by mass) of the interstellar medium, their effect on the flow of radiation in the Galaxy is far greater than that of the gas. They absorb much more light, and emit far more infrared power. They play major roles in the heating of interstellar clouds, and in the formation of new stars. They assist in the synthesis of molecules and in their subsequent protection from ultraviolet radiation. They provided the material out of which the Earth is made. They are also among the most frustrating objects in astronomy, both for what they hide of the Universe and for what they fail to reveal about their own nature.

Although gas and dust are generally fairly well mixed together, they behave in sufficiently different ways that their effects are easy to distinguish. In this chapter we will consider the dust grains in isolation, leaving their interactions with the interstellar gas for later in the book.

7.1 Extinction and reddening

The interstellar medium is nowhere completely transparent and nowhere completely opaque; a dusty region of space dims rather than blocks the light passing through it. This dimming process is called **interstellar extinction**. The amount of extinction in front of a star can be expressed quantitatively, and is usually given the symbol A_λ, where A_λ is the number of magnitudes by which the star's light is dimmed by the dust at the wavelength λ. The **magnitude** system for measuring the brightness of stars is described in Appendix K; a star behind a dust cloud with $A_\lambda = 2$ would appear two magnitudes (a factor of 6.3) fainter than if the dust cloud had not intervened. The extinction due to the interstellar dust varies greatly from place to place. In the plane of the Galaxy the extinction at visible wavelengths (A_v) is typically one magnitude (a factor of 2.5) for every 500 parsecs the light travels through it.

Interstellar dust has a devastating effect on visible light that is trying to cross the Galaxy. Because of the compounding effect of extinction, light is dimmed by a factor of $(2.512)^4$, or nearly 40, in a distance of 2 kpc. Doubling this distance to 4 kpc causes a dimming of 40^2 or 1600. Light that has to cross the 30 000 parsec diameter of the Galaxy is dimmed by 2.51^{60}, a factor of 10^{24}. With this kind of extinction visible light stands no chance, and not even the most powerful telescopes can see to the other side of the Galaxy disk. This is why it is so difficult to study anything other than the local few kiloparsecs of our Galaxy by visible light.

The amount by which light is dimmed by the interstellar medium depends on its wavelength. Red light suffers less extinction than blue or violet light, with the result that the apparent color of a star is altered if it is observed through a patch of dust. This phenomenon is called **interstellar reddening**, and is a consequence of the fact that the particles causing the extinction are smaller than the wavelength of light. A similar kind of reddening occurs in the Earth's atmosphere, and explains why the color of the Sun appears to change at different times of the day. The reddening effect of the Earth's atmosphere is small when the Sun is high in the sky; in the evening, however, when sunlight has to travel a long distance through the atmosphere at a shallow angle, the reddening effect becomes more pronounced and the setting Sun appears orange rather than white.

The fact that dust reddens starlight turns out to be extremely fortunate, because it means that by measuring the color of a star an astronomer can estimate how much dust lies in front of it. If dust caused equal extinction at all wavelengths it would be extremely difficult to know if a faint star was faint because of its great distance or because of dust. The first step towards measuring the extinction in front of a star is to record its spectrum. By checking which elements are ionized and which are neutral an astronomer can estimate the star's surface temperature. From the temperature, the intrinsic color can be estimated using black-body curves like those in Figure 3.8. If the star's apparent color is redder than the color that is predicted from its spectrum then dust must be intervening.

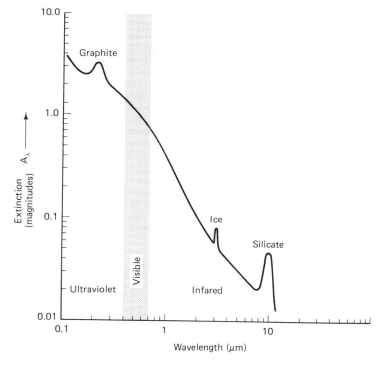

Figure 7.1. An interstellar extinction curve for interstellar dust particles showing how radiation of different wavelengths suffers different amounts of extinction. The curve is compiled from measurements on a variety of stars and is adjusted so that $A_v = 1$ magnitude. The peaks of absorption at 0.22 μm, 3.1 μm and 9.7 μm are probably due to graphite, ice and silicate particles respectively.

Interstellar extinction and reddening are important in the infrared and ultraviolet parts of the spectrum as well as the visible. In this context the term 'reddening' is used, with some chromatic licence, to indicate the relative enhancement of long-wavelength over short-wavelength radiation. Figure 7.1 indicates how the extinction (A_λ) varies with wavelength from

0.1 to 10 μm. The curve shows that at short wavelengths the extinction is several times larger than for visible light, while at infrared wavelengths longward of 2 μm it is reduced by a factor of at least 10.

This enormous improvement in the transparency of the interstellar medium at long wavelengths is of great importance as it allows astronomers to see through layers of dust with $A_v = 20$ magnitudes or more, which would

Figure 7.2. The H^+ region W51 at infrared wavelengths. This image is based on data obtained at 1.2, 1.65, and 2.2 μm. W51 contains a large number of newly-formed OB stars and is one of the most powerful H^+ regions in the Galaxy. It is totally hidden from view at visible wavelengths.

be impenetrable at shorter wavelengths. Figure 7.2 shows an infrared image of the H^+ region W51, which lies about 7 kpc away in the galactic plane. At visible wavelengths the light from W51 is attenuated by 25 magnitudes, or a factor of ten billion (10^{10}), rendering it totally unobservable. At a wavelength of 2.2 μm the extinction drops to 2.5 magnitudes, or a factor of only 10. The attenuation at 2.2 μm is therefore reduced by a factor of *a billion* as compared with the visible. The technique of using infrared telescopes to penetrate thick dust clouds has also been applied with great success for finding newly-formed stars inside dense interstellar clouds (see Chapter 12). An extreme example of the value of making observations at long wavelengths is the use of radio astronomy as the primary tool for determining the structure of distant parts of the Galaxy (Chapter 4). At wavelengths of 1 cm or more no trace of the influence of interstellar dust has ever been detected in any astronomical measurement.

7.2 Scattered light

Two things can happen to light rays when they meet a dust particle; they can either be absorbed onto the grain or scattered back into space. The relative importance of these two processes depends on the size, shape and composition of the grain and on the wavelength of the radiation being scattered. For typical interstellar dust grains at visible wavelengths approximately equal amounts of light are absorbed and scattered. Starlight that has been scattered no longer travels in a straight line from its parent star. When observed from the Earth, such light appears to shine from the patch of interstellar dust where it was last scattered rather than from the star itself. The dust therefore appears to be shining itself.

Scattered starlight makes up about a quarter of the light we see from the Milky Way. Usually the scattered light is too diffuse for either the star or the dust cloud to be identified, but if an interstellar cloud lies close enough to a bright star a **reflection nebula** (which would probably be better named a 'scattering nebula') may be formed. The most famous example of a reflection nebula occurs in the Pleiades cluster of some 3000 young stars in the constellation of Taurus (Figure 3.1). The stars in the cluster are about 120 parsecs away and were all formed as a group about 10^8 years ago. Other examples of reflection nebulae are seen near the boundary of the Ophiuchus and Scorpius constellations (Figure 7.3). The dust clouds of Ophiuchus are among the nearest dense interstellar clouds to the Solar System.

Figure 7.3. Several reflection nebulae are seen in this view of part of the constellations of Ophiuchus and Scorpius. The nebulae at the top and the middle-left surround hot stars, so look blue. The very bright reflection nebula at the bottom left surrounds a cool star (Antares), so appears redder. Several other phenomena are seen in this picture including the pink glow of an H^+ region near the mid-right edge of the picture, dense dark clouds above and to the left of center, and a distant unrelated star cluster at the bottom right of the picture.

Figure 7.4. Electromagnetic waves consist of electrical disturbances that are directed sideways from the direction of the wave's motion. In an unpolarized wave the disturbances occur equally in all directions. In a 100% polarized wave all the disturbances are in one direction, illustrated here as either left–right or up–down. Waves that are polarized left–right in the figure are more efficiently scattered off a horizontal reflecting surface than waves with the opposite polarization. Unpolarized light that is scattered obliquely off the surface will become partially polarized, with more energy in the left–right waves than the up–down waves. 'Polaroid' sunglasses cut down reflected glare by blocking all left–right polarization reaching the eye.

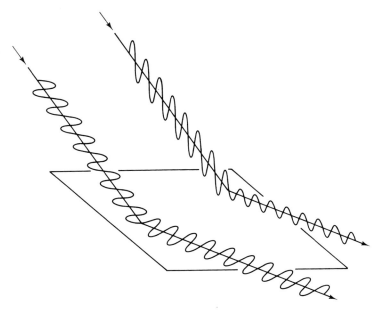

Reflection nebulae are readily distinguishable from H^+ regions by their spectra. Whereas an H^+ region emits its light in a number of bright emission lines, the light from a reflection nebula closely resembles that of the star that illuminates it. The spectrum is continuous with the same absorption lines as the star. The reflection nebula usually has a bluer color than the star, however, because of the greater efficiency with which short wavelength radiation is scattered. The Earth's daytime sky is blue for the same reason; blue light is scattered more than red light.

The stars that illuminate reflection nebulae usually, though not always, have surface temperatures in the range 15 000–30 000 K, considerably hotter than the 5800 K of the Sun. The 30 000 upper limit is set by the fact that stars hotter than this generate enough ultraviolet radiation to ionize the gas associated with the dust. The bright light from the H^+ region that would thereby be formed would greatly outshine the light from the reflection nebula. The lower limit to the temperature is set by the fact that hot stars have short lifetimes and are therefore the ones most likely still to be associated with the remnants of the clouds associated with their births.

Light that has been scattered by dust grains is **polarized**. As an electromagnetic wave travels through space it generates vibrating electrical fields at right angles to its direction of motion. In an *unpolarized* wave these fields have no favored direction, in the sense that they are as much 'up–down' as 'left–right' (Figure 7.4). In a polarized wave the electrical fields are stronger in one direction than in the other. The difference can be measured at the telescope, as can the angle of the preferred direction of polarization. The special filters that are used to measure polarization are similar to those in 'Polaroid' sunglasses.

The direction on the sky where the polarization is a maximum is related to the direction from which the dust is illuminated. This effect can be used to pinpoint the light source in a reflection nebula. The object in Figure 7.5 is part of a group of infrared sources in the constellation Monocerotis. The high degreee of polarization at 2.2 μm and the circular pattern of the polarization leave no room for doubt that what we are seeing is a reflection nebulosity illuminated by a central star.

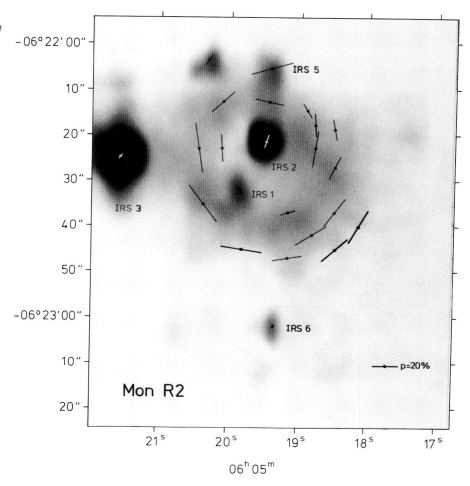

Figure 7.5. The Monoceros R2 nebulosity at a wavelength of 2.2 μm. The dark patches show the areas of strongest infrared emission. The black lines show the direction of polarization of the infrared emission; the circular pattern of the lines indicates that the object marked IRS2 is the source of the illumination of the nebula.

7.3 Infrared emission

Since starlight carries energy, any visible or ultraviolet light that is absorbed (as opposed to scattered) by a dust grain heats it up. The energy the grain absorbs is balanced by the energy it radiates back into space as infrared photons. Heated dust is, in fact, the most widespread source of infrared radiation that astronomers have found. Hot dust has been found within the Solar System, throughout the Galaxy, and in the nuclei of galaxies millions of parsecs away from the Earth.

Like any other solid body, a heated grain radiates a continuum spectrum that depends primarily on its temperature. The temperature is set by the interplay between the input and output of energy. If an object absorbs power faster than it loses it, its temperature rises. As it heats up it becomes able to radiate more power, thereby tending to redress the imbalance. If it has enough time it eventually settles at an equilibrium temperature at which its power input and output are balanced. For an object that absorbs all radiation with 100% efficiency this temperature depends only on the strength of the incoming electromagnetic radiation. An object the size of a planet, and one the size of a pebble would both settle at the same temperature if they were placed at the same distance from a heat source such as a star. These idealized objects are referred to as **black bodies** by physicists. Real objects usually have equilibrium temperatures that differ

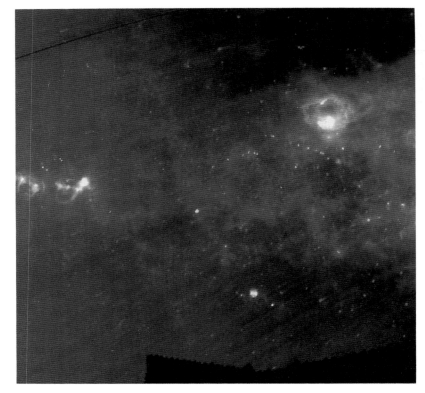

Figure 7.6. Part of the plane of the Milky Way as seen by the IRAS satellite. This image is constructed from scans of the sky made by the spacecraft and show the infrared emission from heated dust. The coolest dust, nicknamed 'cirrus' appears red. Hotter dust associated with nebulae and with newly-formed stars, appears white. The brightest patches are dense clouds heated from within by newly-formed stars.

somewhat from the ideal black-body temperature. The differences may be due to a surface that absorbs and emits better at some wavelengths that others, or to a size that is comparable to or smaller than the wavelength at which most of the energy is emitted. Interstellar dust grains usually fall into the latter category, and have temperatures higher than an idealized black body.

Most interstellar dust grains intercept very little power. They are heated only by the energy of the faint light from myriads of stars in the Milky Way. Under these circumstances they settle at equilibrium temperatures of a few tens of degrees above absolute zero. As predicted by Wien's law (see Section 3.8 and Appendix H), the faint radiation that they emit is strongest at wavelengths of 100 μm or more. The emission from this diffuse interstellar dust was mapped in detail by the Infrared Astronomy Satellite (IRAS) in 1983 (Figure 7.6). The patchy structure of the 100-μm emission from interstellar dust has earned it the name of **infrared cirrus** in recognition of its resemblance to the high-altitude clouds that plague astronomers at ground-based observatories. The infrared data provided much more information about the distribution of interstellar dust than had ever been obtained from studies of interstellar extinction. For one thing IRAS was able to detect 100-μm emission from layers of dust too tenuous to produce measurable reddening: for another it revealed the pattern of dust over the whole sky, and not just in front of a number of stars. The patterns of the cirrus emission correlate well with the patterns of 21-cm emission from neutral hydrogen in the same direction. This result provides evidence that the gas and dust go together in space, and that we are right to consider that an interstellar cloud contains both types of matter.

Only satellite-borne telescopes are able to map the cirrus emission, but if dust grains are raised to higher temperatures they become much easier

Figure 7.7. Infrared emission from dust shells around stars. The bright star μ Cephei emits about 10% of its energy in an infrared excess at 10–20 μm. The more extreme infrared source IRC + 10216 emits essentially all of its energy at wavelengths longer than 3 μm.

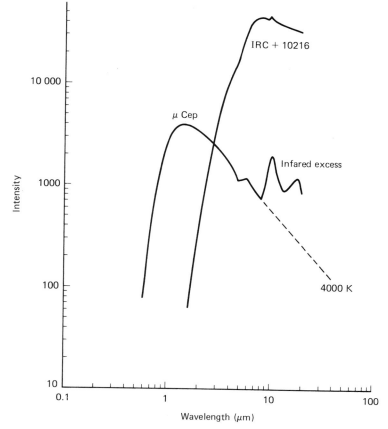

to detect. They radiate more power, but just as importantly, their power is emitted at shorter wavelengths. When dust is hotter than about 150 K it radiates strongly at wavelengths of 20 μm and less, permitting observations with ground-based telescopes. Dust grains rarely get hot enough to emit visible light, however, since most are destroyed by being heated above 1500 K.

Infrared emission from heated grains is most easily observed when the dust grains are in the immediate vicinity of powerful stars. Such a situation commonly occurs around giant stars which are ejecting material from their surfaces out into space. The bright star μ Cephei is an example. Shortward of 7 μm its emission resembles that of a 4000 K black body, as would be expected from the pattern of absorption lines seen in its visible spectrum (Figure 7.7). Instead of continuing to decrease longward of 7 μm, however, the emission increases sharply at 10 μm and again at 20 μm. These are the wavelengths at which heated silicate dust grains radiate most efficiently; the presence of these infrared 'bumps' indicate that there is a **circumstellar shell** of dust surrounding the star. The dust particles absorb about 10% of the light from the star and re-radiate this energy as infrared waves. The **dust shell** is too small to be seen directly, and its presence can be inferred only from its infrared radiation. A more extreme example of a star surrounded by a dust shell is provided by the object IRC + 10216, a star with a surface temperature of 2000 K. In this case essentially all of the light from the star is absorbed and re-radiated at infrared wavelengths. Although this star is extremely faint at visible wavelengths, it is one of the ten brightest stars in the sky at a wavelength of 20 μm. Most of the infrared flux comes from a shell of gas and dust 0.4 arcseconds in diameter at a temperature of 600 K. The distance to IRC +10216 may be about 200 light years. If so, then the dust cloud has a diameter of 80 astronomical units (the same as Pluto's orbit) and a power output 25 000 times that of the Sun. The role that giant stars play in resupplying gas and dust to the interstellar medium is discussed in Chapter 10.

Another common circumstance where dust and stars occur close enough together to make strong infrared emission is near the birth places of stars. As we shall discuss in Chapter 12, the clouds within which stars first form are generally so dense that no visible light can penetrate out from their interiors. A newly-formed star is therefore often first detected by the infrared emission from dust in its immediate surroundings. Depending on the progress of the star's birth this dust may be part of the gas cloud that is collapsing to form the star, or may be surplus material that is being ejected by the star back into space.

Inside of an H^+ region dust is heated by a two-stage process in which the energy from the star is first passed to the gas and then to the dust. As we discussed in Chapter 5, when hydrogen ions and electrons recombine in an H^+ region they produce emission lines. The strongest of these recombination lines are the transitions from various upper levels all the way down to the $n=1$ ground states. They have wavelengths in the range 912 to 1216 Å. These ultraviolet photons are repeatedly scattered by encounters with neutral atoms and become trapped inside the H^+ region. This trapping process greatly increases the probability that these photons will get absorbed by dust grains before they can escape into space. By this cooperative process between gas and dust, at least a third of the power from an OB star gets converted to infrared radiation inside the ionized region itself. Dust grains mixed with neutral gas just outside of the H^+ region may raise this fraction further. Given that OB stars are always highly luminous, it is not surprising that H^+ regions are the most powerful sources of infrared radiation in the Galaxy.

7.4 Tools of the trade: infrared astronomy

The first experiments to measure infrared emission from celestial objects were made in the nineteenth century, but infrared observations did not start to have a major impact on astronomy until the late 1960s. Infrared astronomers are faced with three major technical problems over and above those which optical astronomers have to solve.

Their first problem is that the Earth's atmosphere is opaque at most infrared wavelengths. Only radiation within certain restricted wavelength ranges – called **atmospheric windows** – can reach the ground. Among the more important of these windows are those between 2.0 and 2.4 μm, and between 8 and 13 μm. Radiation in these wavelength ranges, plus a few others, reaches the ground more or less intact, but radiation outside of these windows is heavily absorbed by the gases in the Earth's atmosphere. The 'windows' get murkier at longer wavelengths. Between 18 and 35 μm the transparency depends crucially on the amount of water vapor in the atmosphere; for this reason infrared telescopes are usually located at high altitudes where the air is thin and dry. Both of the world's largest infrared telescopes – the 3.8-m United Kingdom Infrared Telescope (UKIRT) and the 3-m NASA Infrared Telescope Facility (IRTF) – are situated on the peak of Mauna Kea, a 4000-m (14 000 ft) volcanic peak in Hawaii. Infrared telescopes for the 0.7 to 35 μm wavelength range look rather similar to optical telescopes and, indeed, can be interchanged with them for some kinds of observations.

At wavelengths longer than about 35 μm essentially all infrared radiation is absorbed in the Earth's atmosphere. Observations must be made either from space or from telescopes mounted in high flying aircraft. The **Kuiper Airborne Observatory** (KAO) is a facility operated by NASA that allows research astronomers to catch infrared photons before they are absorbed in the lower levels of the Earth's atmosphere (Figure 7.8). Two or three nights a week it carries teams of astronomers to altitudes of about 13 km (43 000 ft). While the aircraft flies on a steady course well above the clouds and the water vapor the observatory's 91-cm diameter telescope is pointed at the sky through a large hole in the side of the plane. Plans are under consideration for a 3-m diameter flying observatory carried in a converted Boeing 747SP aircraft.

Figure 7.8. The Kuiper Airborne Observatory consists of a 91-cm diameter infrared telescope mounted inside a Lockheed C-141 transport aircraft. The telescope looks out through the square hole in front of the wing. The observatory allows astronomers to observe interstellar matter at wavelengths that are strongly absorbed by the Earth's atmosphere.

Among satellites designed for infrared astronomy, the most important has been **IRAS** – the Infrared Astronomy Satellite. A joint project of the USA, the Netherlands and the United Kingdom, IRAS orbited the Earth in 1983, observing more than $\frac{3}{4}$ million stars, asteroids and galaxies in its one-year lifetime. IRAS was built to make a survey of the sky in the 8–120 μm wavelength range. It will take years of painstaking follow-up work from the ground and from the Kuiper observatory to understand the nature of all the objects IRAS detected. The Cosmic Background Explorer satellite (**COBE**) was launched by NASA in 1989 to map the large scale pattern of infrared emission from the whole sky. Both the USA and Europe have plans for advanced remote-controlled space telescopes designed to observe specific sources of infrared radiation in detail.

The second problem that infrared astronomers have to face is that everything on Earth, including telescopes, the atmosphere and even the astronomers themselves are strong natural sources of thermal infrared radiation. The thermal radiation produced by the mirror of the telescope is generally much stronger than the astronomical signal being measured, and enormous care must be taken to minimize the effect of spurious signals such as these. Since the thermal radiation from an object depends strongly on its temperature, one way of minimizing the emission from the telescope is to refrigerate it. This approach was used in the IRAS satellite, the main mirror of which was cooled by liquid helium to a temperature of around 10 K. Unfortunately this approach cannot be used for telescopes in ground-based observatories, since the gases in the Earth's atmosphere would condense on the mirror and totally ruin its optical performance.

The third problem that infrared astronomers have been faced with is the need for sensitive detectors. Photographic plates can be used only between 0.3 and 0.9 μm, and are no good over the bulk of the infrared waveband. Until the late 1980s infrared astronomers were forced to make most of their observations using single element electronic detectors. To make a map of a galaxy an astronomer would have to measure the signal from each point in the sky separately – an extremely time-consuming process. In the past few years, however, arrays of multiple infrared detectors have become available, with up to 60 000 separate detectors arranged in a square pattern. These infrared array cameras are revolutionizing infrared astronomy, vastly increasing the rate at which data can be gathered. The greatest gains so far have been at wavelengths near 2 μm. Among professional astronomers, observations at these wavelengths have now become almost as routine as those at visible wavelengths.

7.5 Sizes and shapes of dust particles

The extinction curve (Figure 7.1) provides us with what little knowledge we have about the sizes of dust particles. Theory tells us that when a very small obstacle is put in the way of an electromagnetic wave the amount of energy scattered or absorbed depends strongly on the wavelength. With a larger obstacle the dependence on wavelength is weaker. In other words small grains cause more color change (i.e. reddening) than large grains. It is found that the extinction curve in the visible part of the electromagnetic spectrum agrees fairly well with theory if the dust particles are assumed to be tiny spheres each with a diameter of about 0.1 μm. To explain the infrared and ultraviolet parts of the extinction curve we need additional grains that are both larger and smaller than 0.1 μm. The ultraviolet extinction comes from large numbers of smaller particles with sizes going down to 0.005 μm, while the infrared extinction is caused by a smaller number of particles ranging up to 1 μm in size. There could be additional dust grains outside these size limits but we would need more data at longer and shorter wavelengths in order to see their effects strongly.

Individual dust grains are far smaller than anything we can see with our eyes. They are more like smoke than sand. Each 0.1 μm interstellar grain weighs only 10^{-15} grams; 5 000 of them could be lined up across the full stop at the end of this sentence. Small though they are, they each contain millions of atoms.

Our knowledge of the shapes of grains is even skimpier than our knowledge of their sizes. For simplicity, astronomers often assume that they are roughly spherical, but grains in the form of long thin fibers and in thin flat sheets have also been suggested. The only thing that we know for certain is that at least some grains are not spherical. The evidence comes from observations of **interstellar polarization**. When the light from a reddened star is examined carefully it is generally found to be slightly polarized. The amount of polarization is greatest for stars which are the most reddened, so it is a natural conclusion that the polarization arises during the passage of the starlight through the interstellar medium. At visible wavelengths the measured polarization is usually something like 1% for every magnitude of visual extinction (A_v). The existence of this polarization implies that light waves vibrating in some directions get absorbed more strongly than light vibrating in other directions. There is no way that such a difference can be produced by grains which are perfectly spherical, so we may conclude that irregular, probably elongated, grains are responsible. As we shall discuss in Chapter 9, these grains are aligned parallel to each other by forces produced by the **interstellar magnetic field**. Note that the polarization of starlight that we are discussing here is a different phenomenon to the polarization that is seen when we look at light from a reflection nebula. Light that has been scattered may be polarized by grains of any shape, but light we view from a star itself can only be polarized by non-spherical grains.

7.6 The nature of interstellar grains

To determine what grains are made of we must turn to spectroscopy. Unfortunately, spectroscopy is only of very limited value when it comes to analyzing dust. Unlike gases, solids produce no sharp spectral lines by which they may be confidently identified. When solids absorb and emit radiation they do so over broad bands of wavelengths. A few of these bands can be picked out by spectroscopy but mostly they overlap each other to produce the rather smooth general extinction curve that is seen in Figure 7.1, or the infrared emission curves of Figure 7.7. At least six major spectral features, or groups of spectral features, have so far been found. Most of these are at infrared wavelengths. Generally, these features have not been identified with specific substances, only with groups of materials with some chemical similarities.

(1) First, there is the 2200-Å (0.22-μm) feature. This is the only dust feature in the ultraviolet region. Because it is so strong it has to be caused by a substance which is abundant in the interstellar medium. The most likely candidate is **graphite**, one of the commonest forms of the element carbon, and the main ingredient of pencil leads.

(2) The strongest infrared absorption feature is centered at 9.7 μm, but is so broad it affects all wavelengths between 8 and 12 μm. The absorption is attributed to **silicate** particles. These are chemical compounds based around a grouping of silicon and oxygen atoms in conjunction with other common elements such as iron, aluminum, calcium and magnesium. A vast number of different types of silicate materials are possible, many of which are found as minerals on the Earth and Moon. The shape of the absorption feature does not match that of any known mineral, however, leading most astronomers to suspect that it is caused by a combination of many different types of silicate particles with slightly different absorptions. A weaker absorption feature at about 20 μm is also attributed to silicate particles. If silicate dust grains are heated, the 9.7 μm and 20 μm bands can appear in emission, giving rise to spectra like that of the star μ Cephei (Figure 7.7).

(3) Less certain is the identity of a group of weak absorption features near 3.4 μm and 6.8 μm. The wavelengths of those absorptions correspond to groupings involving carbon atoms, such as —CH_2— and =CO. It is speculated that these grains are **organic refractories** of some sort. That is to say they are made from carbon-containing molecules that do not evaporate easily. Possibly they are some kind of polymer. The prominent British astronomer Fred Hoyle has made the point that the absorption features resemble those of certain bacteria, and has cited this similarity as evidence for the idea that life originated elsewhere in

the Galaxy rather than on Earth. Hoyle's interpretation has not been widely accepted by astronomers.

(4) Spectra of infrared sources that are hidden deep within dense interstellar clouds sometimes show strong absorption features attributable to small frozen molecules. The most prominent features at 3.1 and 6.0 μm are probably due to water (H_2O) ice. Evidence for frozen carbon monoxide (CO), and possibly ammonia (NH_3), and methyl alcohol (CH_3OH) has also been found. Because these four chemicals all sublime into gases at low temperatures they cannot survive in solid form away from the protection offered them by the dark interior of an interstellar cloud. One theory is that inside the clouds these ice-like substances may condense as coatings on the outside of the silicate and graphite grains.

(5) The least understood absorption features are a series of over 50 **diffuse interstellar bands** in the visible spectra of reddened stars. The strongest are at 4430, 6284 and 6177 Å. Whereas an atomic absorption line is typically less than 0.2 Å wide, the interstellar bands are usually 10 to 100 times wider – far too large to be explained by Doppler broadening. In the visible wavelength range the diffuse lines absorb about six times more light than do the atomic and ionic lines. The absorption is still less than a thousandth of that produced by the dust, however. Because the diffuse interstellar bands are strongest in regions of high extinction it is assumed that they are associated with dust grains in some way, but as yet astronomers have had virtually no success in identifying the materials responsible.

(6) Finally there are strong infrared emission features at 3.3, 6.2, 7.7, 8.6 and 11.3 μm (Figure 7.9). These features differ in two important ways from the others discussed in this section. First, they are never seen in absorption. Second, they are only seen in situations where dust is being heated by ultraviolet radiation, such as in H^+ regions or planetary nebulae. The features have been tentatively identified with a group of molecules called **polycyclic aromatic hydrocarbons** or PAHs for short. They consist of 10–50 carbon atoms bonded together in stable rings, with hydrogen atoms around the edges. As a group, the molecules are remarkably stable, and are strong enough to survive in interstellar space. When one of these molecules absorbs an ultraviolet photon it is raised to an excited state from which it drops back to the ground level in a series of jumps, emitting infrared photons as it does so. These photons produce the emission bands seen in Figure 7.9. No single molecule can account for all the infrared bands seen, so it is assumed that a variety of related molecules is present.

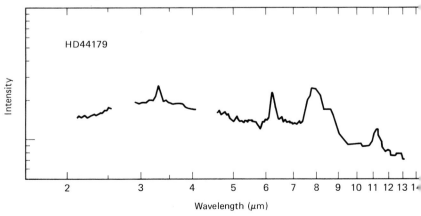

Figure 7.9. Infrared spectrum of the small nebula HD 44179. The emission bands are believed to come from minute carbon-bearing molecules called polycyclic aromatic hydrocarbons (PAHs).

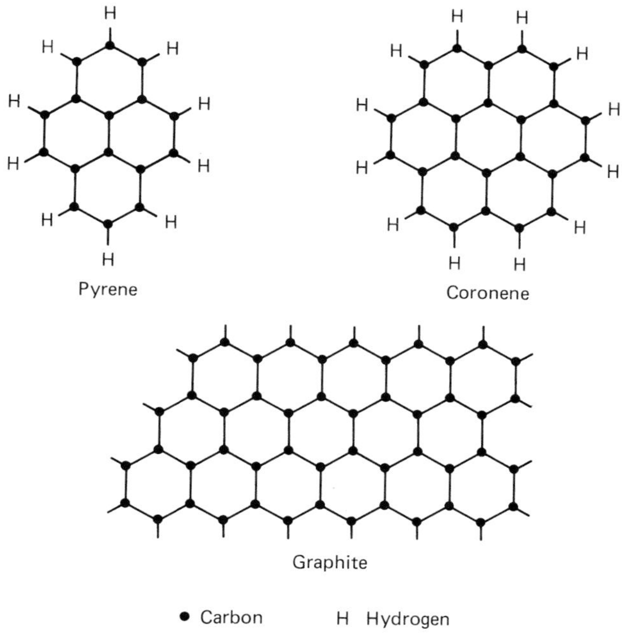

Figure 7.10. PAH molecules (such as pyrene or coronene) and graphite grains are both comprised of hexagonal rings of carbon molecules.

Pyrene

Coronene

Graphite

• Carbon   H Hydrogen

Is it fair to describe a structure with less than 50 atoms as a dust grain rather than a molecule? Probably not. An interesting point about PAHs, however, is their underlying resemblance to an individual layer of graphite (Figure 7.10), which, give or take a few hydrogen atoms round the edge, consists of identical, though far more numerous, carbon rings. Could there be 'particles' with a number of carbon atoms intermediate between that of a PAH molecule and of a normal graphite grain? Some tantalizing evidence exists that the answer is yes. In 1983, astronomers searching in a reflection nebula found continuous infrared emission that appeared to be coming from grains that were far hotter than could be explained by steady heating from the stars in the area. What seems to be happening is that there are grains in the nebula so small that an individual photon of starlight contains enough energy to raise its temperature to 1000 K. It maintains this temperature for only an instant before it cools back to its normal lower equilibrium temperature. Evidence for these minute grains has since been found in the general interstellar medium, and in the nuclear regions of other galaxies.

One of the few strong conclusions one can draw from the spectroscopic data is that there are a number of different kinds of dust particle. There are silicates, graphite, organic compounds and ices, at the very least. Quite probably there are other substances for which we as yet have no direct evidence. Silicon carbide and magnesium oxide have been found in the outer atmospheres of certain stars, and it is a good bet that these substances could be found in the interstellar medium if our equipment could be made sensitive enough. Scientists who try to come up with a 'recipe' for interstellar dust do not rely solely on spectroscopy. They incorporate data on the abundances of different dust-forming elements and on the likely method of production of each type of grain. They also pay much attention to the way certain elements are depleted from the gas phase of the interstellar medium. Fitting all these diverse data is hard and no unambiguous conclusions can yet be drawn. There is a tendency for dust recipes to change in major ways with each new spectroscopic discovery. Probably the only statement that it is reasonable to make is that in the general interstellar medium there are very roughly equal masses of silicate-based and carbon-based grains, and that in molecular clouds there is a roughly equal additional mass of ice and other frozen volatiles.

7.7 How much dust is there?

If we know what a grain is made from and how big it is, we can calculate how much extinction it produces. From this we can estimate how densely packed in space the grains have to be in order to produce the extinction we see. What we find is that most dust grains live lonely lives. In a typical region of interstellar space, where there is one gas atom cm^{-3}, dust grains are separated from each other by distances of tens or hundreds of meters. They rarely come into contact with another although collisions between the gas and the dust are much more common.

The extinction in any direction can be compared with the amount of interstellar gas in the same line of sight. The latter can be obtained either from the 21-cm line, or from measurements of ultraviolet absorption lines of atomic and molcular hydrogen. The current best estimates put the ratio of dust to gas in the general interstellar medium as 0.7% by mass. In other words, there is 140 times more gas (mainly hydrogen and helium) than there is dust. This ratio is extremely interesting in light of what we know about the 'cosmic' abundances of the elements. As we have seen, hydrogen and helium make up 99% of the mass of the Universe. They are almost no use for making grains, however, because neither is a stable solid at the temperatures and densities of interstellar space. Hydrogen can become part of a solid in combination with some other common elements, but even at its most efficient (the formation of solid methane with a formula CH_4) this process can produce a dust grain containing only 25% (by weight) of hydrogen. We are therefore forced to conclude that a very significant fraction (of the order of half) of the heavy elements in the interstellar medium are locked up in the form of dust grains rather than gases. It is the participation by these elements in grain-building that is the cause of their depletions from the gas phase of the interstellar medium (Chapter 6). It is also the reason why scientists are fairly sure that the bulk of the grains have to be made mainly from elements that are cosmically abundant like oxygen and carbon.

Finally, in this pollution-conscious age, we might note that the interstellar medium is one of the dirtiest samples of gas that one could find outside of a smokestack. Even the worst urban smog has nothing like the fraction of particulate matter that is found in space. If the Earth's atmosphere contained 0.7% by mass of interstellar-like dust grains it would be so opaque that one would not be able to see as far as the tip of one's nose.

8 Molecular clouds

Before about 1970 there was little evidence that molecules existed in significant numbers in the interstellar medium. Optical astronomers had detected a few weak absorption lines of the CN, CH and CH$^+$ molecules, radio astronomers had discovered maser-like emission lines from OH and H$_2$O molecules, and theoreticians had predicted that H$_2$ molecules ought to exist in dark clouds. Few were prepared for the deluge of discoveries that ensued from the opening up of the ultraviolet, infrared and millimeter wavebands during the 1970s. Dozens of new molecules were found in rapid succession, but the most fundamental breakthrough was the discovery of the **giant molecular clouds**. These vast accumulations of hydrogen molecules and other more complicated chemical compounds are the most massive objects in the Galaxy.

8.1 Molecular spectroscopy

A molecule is a much more complicated object than an atom, and consequently has many more energy levels available to it. The number of possible spectral transitions is correspondingly greater, particularly at infrared and microwave wavelengths. The richness makes molecular spectroscopy much more complex than atomic spectroscopy, but it potentially makes an enormous amount of astrophysical information available to astronomers.

The transitions of a simple molecule such as carbon monoxide (CO) can be divided into three main types: electronic, vibrational and rotational. **Electronic transitions** involve changes in the shape of the cloud of electrons surrounding the constituent atoms. They are very similar to the transitions seen in single atoms, and, like them, occur mainly at ultraviolet wavelengths.

Vibrational transitions occur as a result of elasticities within the molecule that allow its atoms to oscillate to and fro with respect to each other. Because these oscillations are governed by the laws of quantum mechanics the vibrational energy of a molecule can have only certain values. In a vibrational transition a molecule switches from one vibrational energy level to another, but does not change its electron configuration. As with all kinds of transition, the energy difference may be made up by a photon or from the kinetic energy of a colliding particle. Because vibrational transitions usually involve smaller energy changes than electronic transitions the photons emitted or absorbed have longer wavelength. They are usually found in the infrared waveband. A carbon monoxide (CO) molecule in its

80 MOLECULAR CLOUDS

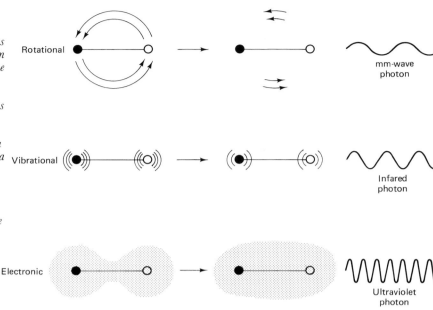

Figure 8.1. The three most important types of transition that a simple molecule such as carbon monoxide can make. In a rotational transition only the speed of its end-over-end spinning is changed. A vibrational transition includes a change in the amplitude of the to-and-fro motions of its atoms, and usually involves a larger energy difference than a pure rotational transition. The transitions with the largest energy differences are the electronic transitions, which include a change in the shape of the electron cloud. Rotational, vibrational and electronic transitions most often occur at millimeter, infrared and ultraviolet wavelengths respectively.

electronic ground state requires a photon of wavelength 4.6 μm to jump from the lowest to the next lowest vibrational energy level.

Rotational transitions result from changes in the rate at which a molecule spins. Once again, the energy of a rotating molecule is limited by quantum mechanics to certain fixed values, so that each electronic and vibrational energy state is split into a number of rotational energy levels. These energy levels are closer together than electronic or vibrational energy levels so the photon energies involved in rotational transitions are smaller. They produce spectral lines which are usually at microwave or millimeter wavelengths; the simplest rotational transition of the CO molecule has a wavelength of 2.6 mm.

There are several complications to this simple picture that we must be aware of. First, a vibrational transition often triggers a simultaneous change in the molecule's rotation, while an electronic transition is likely to alter both its vibration and rotation states. Second, molecules with more than two atoms can vibrate in several ways simultaneously, and, if their atoms do not all lie in a straight line, will spin at different rates around different axes. Third, molecules which have an odd number of electrons in them have each of their rotational energy levels subdivided by a phenomenon called **hyperfine splitting**, which arises because of magnetic effects due to the 'spare' electron. Although these complications tend to make life difficult for those trying to understand the physics of molecules, they can be a boon for astronomers because they increase the number of potentially observable spectral lines. Of particular value are processes which produce very small differences between energy levels, because the spectral lines thus produced have long wavelengths that are easy to observe with radio telescopes. The earliest molecular lines observed by astronomers, such as those of hydroxyl (OH), water vapor (H_2O), ammonia (NH_3), and formaldehyde (H_2CO), all had wavelengths longer than 1 cm and made use of some kind of subtle effect in molecular physics. Since the development of millimeter astronomy in the 1970s, however, the most commonly observed molecular spectral lines have been simple rotational transitions.

8.2 Hydrogen molecules in transparent clouds

Unfortunately, the most common interstellar molecule of all, H_2, has no transitions at millimeter wavelengths and must be observed by special techniques. Hydrogen molecules in space are best observed by the ultraviolet absorption lines they produce in the spectra of stars lying behind interstellar clouds. The absorptions are electronic transitions, but they all also involve simultaneous changes of the molecules' vibrational and rotational states. Most of the studies of molecular hydrogen in space were made by the Copernicus satellite in the 1970s. Since the Copernicus observations were made by analyzing ultraviolet light from a star after it had passed through the intervening interstellar gas only relatively transparent regions of space can be analyzed by this technique.

A great deal of information can be determined from the ultraviolet absorption lines of molecular hydrogen. First, by comparing their strengths with those of atomic hydrogen (such as the 1216-Å Lyman-α line) we can find out what fraction of the hydrogen in any given direction is in molecular as opposed to atomic form. The principles are the same as those used to determine the abundances of the elements. The results turn out to be extremely interesting, particularly when we compare the amount of molecular hydrogen with the amount of interstellar extinction in the lines of sight to different stars. We find that, if there is not much dust in front of a star, very little of the hydrogen in that direction is molecular. However, if there is an interstellar cloud in front of the star which dims it significantly we find that the hydrogen is almost all molecular. The implication of this result is that hydrogen molecules are common inside interstellar clouds, but are rare outside them. We can regard this result as an extension of the idea of interstellar phases. In the same way that hydrogen is found nearly everywhere to be either almost fully ionized or almost completely neutral, we now find that the neutral gas itself is generally almost entirely atomic or almost entirely molecular.

To explain why molecules predominate in some regions but not in others we must look at the factors that affect their formation and destruction. To make a molecule we have to bring its constituent atoms close enough together that they can react chemically. One way of encouraging this process is by increasing the density of the gas, so that collisions between particles occur more frequently. So long as the collisions are reasonably gentle, as they are under most interstellar conditions, high gas densities favor the formation of molecules. Once formed, a molecule is susceptible to several destructive influences. Chief among these are ultraviolet photons which, if they have enough energy, can **dissociate** the molecule into its constituent atoms. The process of dissociation is quite similar to that of ionization, except that the resulting particles are now neutral atoms rather than charged ions and electrons. Less energy is required to dissociate a hydrogen molecule than to ionize a hydrogen atom, and hydrogen molecules are at risk whenever they are exposed to the general starlight background of the Milky Way. The best chance they have for survival is to be protected from this starlight by interstellar dust grains which, as we have seen, are very good at absorbing ultraviolet radiation. The two factors that favor the production and preservation of molecules, namely high gas density and a protective layer of dust grains, are much better met inside interstellar clouds than in the diffuse interstellar medium. It is for these reasons that the neutral interstellar medium is generally either mainly atomic or mainly molecular.

There is a second very important piece of information that we can learn from the ultraviolet absorption lines of molecular hydrogen. They provide us with an excellent measure of the gas temperature. The principle behind this technique is the idea of **statistical equilibrium**. Under normal interstellar conditions hydrogen molecules revert quickly to their lowest electronic state and to their lowest vibrational state by a series of rapid spontaneous transitions. Once in this lowest electronic/vibrational state, however, life is more complicated, since spontaneous transitions among different rotational states take place only extremely rarely. Left to itself, a hydrogen molecule would take more than a thousand years to radiate away the necessary photons to drop into its lowest rotational energy state. In an interstellar cloud, however, molecules are not left to themselves. Collisional transitions between the levels take place, and an equilibrium is soon set up so that the rate at which molecules are excited into any particular level is matched by the rate at which they are excited out again. This equilibrium depends strongly on the temperature of the gas for the following reason: any collision can trigger a downward transition, but because upward transitions require the input of kinetic energy they can be triggered only by collisions more violent than some threshold value. In other words, for any particular pair of energy levels the ratio of upward to downward collisional transitions depends on the fraction of molecules which have more kinetic energy than some value. This fraction can be calculated by applying statistical ideas to physical laws, with a startlingly simple result; so long as collisional transitions dominate over spontaneous transitions the distribution of molecules among the different energy levels depends only on the gas temperature. Under these conditions the relative populations of the different levels are given by what physicists call the **Boltzmann equation** (Appendix L). This equation describes how, as the temperature of a gas increases, a greater fraction of its molecules will be found in excited states, and that levels which require a lot of energy to be excited are generally less utilized than those nearer the ground state. The real importance of the Boltzmann equation is that it puts the connection between temperature and the energy levels in strict quantitative terms.

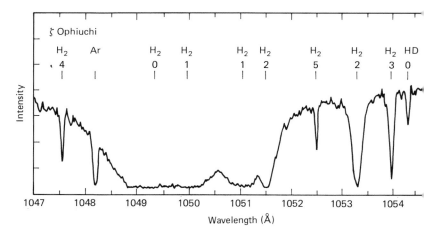

Figure 8.2. Ultraviolet absorption lines of H_2 in front of the star ζ Ophiuchi as seen by the Copernicus satellite. The number above each line represents the rotation state of the molecule before absorption. Note that several of the transitions are strongly saturated, and that the absorption lines are strongest for low rotation numbers. The temperature of the gas in transparent molecular clouds can be determined from measurements of the relative strengths of these absorption lines. The line marked Ar arises from neutral argon atoms. The line marked HD is caused by hydrogen molecules in which one of the hydrogen atoms contains an extra neutron. The D stands for deuterium, the name given to this heavy isotope of hydrogen.

The way we apply the Boltzmann equation to ultraviolet spectroscopy is as follows. First we must identify each hydrogen absorption line to determine what the rotational energy of the molecule was *before* the absorption took place. This we can do by comparing its wavelength with those measured previously in a laboratory. Then we use the measured strengths of different absorption lines to calculate the number of molecules in each rotation level prior to absorption. The final step is to use the Boltzmann equation to figure the temperature from the relative populations. This technique has been

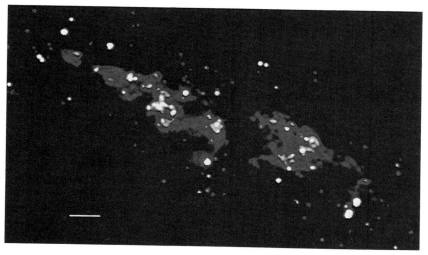

Figure 8.3. Infrared image of vibrationally-excited hydrogen molecules around a newly-formed star. In this region, known to astronomers as DR21, gas that has been ejected from the star collides with the surrounding interstellar gas. In the resulting shock waves hydrogen molecules are excited into an upper vibrational energy level. The radiation seen in this picture is produced as the hydrogen molecules revert to the ground state. DR21 is an example of a bipolar flow (Chapter 12).

applied to clouds in the line-of-sight to a number of stars. The temperatures derived have been in the range 45–125 K, with a mean of 77 K. These temperatures are very similar to those measured for the *atomic* gas in what we referred to as 'cool' clouds in Chapter 4. Those measurements employed the 21-cm absorption line. The reasons why the atomic and molecular gas should have this temperature will be explored in Chapter 11.

8.3 Vibrationally excited hydrogen molecules

Because ultraviolet radiation is so easily absorbed by dust, only relatively transparent regions of space can be studied by ultraviolet spectroscopy. To probe the denser, more opaque regions we need to use transitions in the infrared or microwave regions of the spectrum. We might consider looking for absorption lines in the spectrum of an infrared source situated behind a dense molecular cloud. Hydrogen absorption lines could in principle be seen as a result of molecules being excited out of their ground state to a level of higher vibrational excitation. Unfortunately, hydrogen molecules change their vibration state only reluctantly, and these transitions are too weak to have ever been seen in absorption. In 1976, however, they were unexpectedly discovered *in emission* at around 2 μm wavelength from within a dense molecular cloud behind the Orion Nebula. They have since been seen in several other locations in the Galaxy. The strange thing about these emission lines, and the reason they were not anticipated, is that the relative strengths of the lines suggests that they are generated in a region of the cloud where the temperature seems to be about 2000 K.

What excites the hydrogen molecules in these regions? There are two possibilities. The explanation that seems to apply best in most cases is that the infrared emission lines come from a region where two streams of gas are colliding at a relative velocity of around 20 km s^{-1}. This velocity difference is greater than the speed of sound in the gas. As a result, shock waves are generated in which the streaming motions of the moving gas streams are converted into random thermal motions. The gas in the shock waves is briefly heated to temperatures of around 2000 K, and as a result of the ensuing energetic collisions some hydrogen molecules are excited into upper vibrational levels. As the gas cools back to its original temperature of around 50 K the molecules revert to the lowest vibrational level by emitting infrared photons.

The other way that the infrared emission lines of hydrogen can be generated is by **ultraviolet fluorescence**. This process can work if there is a hot star in the vicinity emitting ultraviolet radiation. The ultraviolet photons are absorbed by the hydrogen molecules, putting them into an excited state via an electronic transition. The excited molecules then revert spontaneously to the lowest electronic state, but when they do this they do not always drop all the way to the lowest vibrational state. Some end up in one of the excited vibrational states, and then have to emit an infrared photon to get back to the lowest vibrational level. If there is a lot of dust associated with the hydrogen the infrared photons may be observable even if the ultraviolet ones are not.

Infrared emission from molecular hydrogen is thus only seen under very special circumstances of shock wave heating or ultraviolet fluorescence. At the normal temperatures found in interstellar clouds, 10–100 K, the infrared lines are not seen at all. If we want to study normal gas in these clouds we need to find spectral lines that result from transitions which can be excited by the comparatively gentle collisions that occur at these low temperatures. Since low energy necessarily implies long wavelength we must look for transitions that have millimeter or radio wavelengths. Unfortunately, hydrogen molecules do not have any spectrum lines in this wavelength range. To study dense (as opposed to transparent) molecular clouds we are therefore forced to observe molecules which *do* have spectral lines at convenient wavelengths. The molecule that is used most often is carbon monoxide (CO). In order to understand how we can interpret the data from this and other molecules in space we must turn our attention to the chemistry of interstellar clouds.

8.4 Interstellar chemistry

Table 8.1 is a list of all the molecules discovered in space through 1987. Most were detected as a result of millimeter-wave spectroscopy of dense molecular clouds in the direction of the constellations Sagittarius, Orion and Taurus. They range in complexity from simple two-atom molecules like carbon sulfide (CS) to molecules with nine or more atoms like ethyl alcohol (C_2H_5OH). As on Earth most, though not all, of the molecules are organic. This means they contain carbon atoms, but does not imply biological origin. Most are electrically neutral, but a few positively charged **molecular ions** have also been detected. Some molecules are familiar to us, like ammonia (NH_3), and carbon monoxide (CO). Others, such as the molecules HC_7N and $HC_{11}N$ are rarely found on Earth, even in specialized chemistry laboratories, and do not have names in common use. Explaining why certain molecules are found in space while others are undetectable requires an alliance between the sciences of astronomy and chemistry.

Astronomers can estimate the abundances of interstellar molecules in a cloud from measurements of the strengths of their emission lines, but there are many pitfalls. The worst problem is that molecular hydrogen itself is unobservable at millimeter wavelengths; there are simply no bright transitions at useful wavelengths. If a cloud is too thick for ultraviolet absorption lines to be measured then the amount of molecular hydrogen must be estimated indirectly. One way of doing this is by using infrared observations

Table 8.1. *Molecules detected in interstellar space*

Simple hydrides, oxides, sulfides, and related molecules:

H_2	CO	NH_3	CS
HCl	SiO	SiH_4	SiS
	H_2O	CH_4	OCS
	SO_2		H_2S
	CC		HNO

Nitriles, acetylene derivatives, and related molecules:

HCN	$HC{\equiv}C{-}CN$	$H_3C{-}C{\equiv}C{-}CN$	$H_3C{-}CH_2{-}CN$
H_3CCN	$H(C{\equiv}C)_2{-}CN$	$H_3C{-}C{\equiv}CH$	$H_2C{=}CH{-}CN$
CCCO	$H(C{\equiv}C)_3{-}CN$	$H_3C{-}(C{\equiv}C)_2{-}H$	$HN{=}C$
$HC{\equiv}CH$	$H(C{\equiv}C)_4{-}CN$	$H_3C{-}(C{\equiv}C)_2{-}CN$	$HN{=}C{=}O$
$H_2C{=}CH_2$	$H(C{\equiv}C)_5{-}CN$		$HN{=}C{=}S$

Aldehydes, alcohols, ethers, ketones, amides, and related molecules:

$H_2C{=}O$	H_3COH	$HO{-}CH{=}O$	H_2CNH
$H_2C{=}S$	H_3CCH_2OH	$H_3C{-}O{-}CH{=}O$	H_3CNH_2
$H_3C{-}CH{=}O$	H_3CSH	$H_3C{-}O{-}CH_3$	H_2NCN
$NH_2{-}CH{=}O$		$H_2C{=}C{=}O$	

Cyclic molecules:

C_3H_2
SiC_2

Ions:

CH^+	HCS^+
H_2D^+	$HCNH^+$
HN_2^+	SO^+
$HOCO^+$	HOC^+
	HCO^+

Radicals:

CH	C_2H	CN	HCO
OH	C_3H	C_3N	NO
	C_4H	NS	SO
	C_5H		

to estimate the amount of dust in a cloud, and then assuming that the ratio of gas to dust in the cloud has a standard value. Despite these difficulties the main results are unambiguous; carbon monoxide is far the most abundant observed molecule after hydrogen, with about one molecule for every 10 000 H_2 molecules. The next most common molecule is believed to be nitrogen (N_2), but since nitrogen does not have any spectral lines at convenient wavelengths, this is only an educated guess. Among those that have been definitely detected in space, formaldehyde (H_2CO) is estimated to be the next most common, despite the fact that it is a thousand times rarer than CO. The great abundance of carbon monoxide and the fact that it has a strong emission line at a wavelength (2.6 mm) that is fairly easy to observe, make the molecule by far the most useful tool that astronomers have for the study of molecular clouds. Almost everything we know about the distribution of molecular clouds in our Galaxy, and about the incidence of

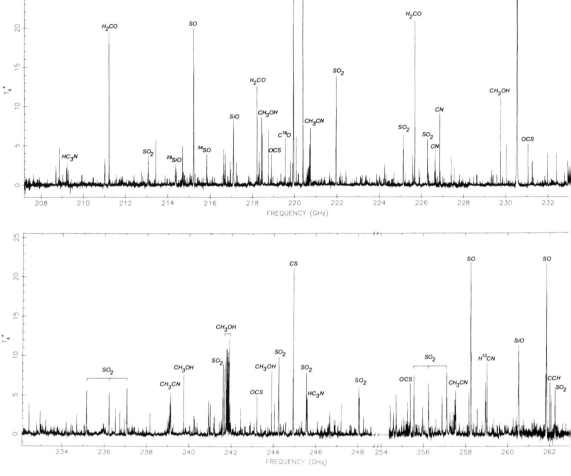

Figure 8.4. Spectrum of a molecular cloud showing emission lines near 1.3 mm wavelength. Nearly 900 lines have been detected, of which all but 13 have been identified with known molecules. The formulae for the molecules producing some of the brighter emission lines are shown.

molecules in other galaxies comes from observations of the CO molecule. The stability of carbon monoxide in space is all the more surprising when we remember that it is regarded as dangerously reactive on Earth, both for its poisonousness and for its flammability.

Why should we find such a different mixture of chemicals in space and on Earth? Most of the differences can be traced back to the fact that many classes of chemical reaction are prevented from occurring in space. Interstellar chemistry is, in theory, much simpler than terrestrial chemistry, although it is much more difficult for scientists to verify their theories experimentally.

Let us look at the factors that make interstellar chemistry different from terrestrial chemistry. First, there are no liquids in space. Reactions between liquids, or between solids and gases dissolved in liquids, comprise the clear majority of chemical reactions which affect our everyday lives. The advantage of liquids for chemistry is that they are denser than gases, but less rigid than solids; opportunities for contact between reacting particles is maximized. Without liquids, the only reactions we need consider in molecular clouds are those taking place between gases, and on the surfaces of interstellar grains.

The second factor that restricts interstellar chemistry is low density. Because particles are far apart in space they encounter each other more rarely. All chemical reactions are therefore slowed down, but some are affected much more than others. Reactions that require three or more gas particles to come together within a short time are common on Earth, but essentially impossible in molecular clouds. While low densities inhibit the formation of some molecules, they can allow the build-up of others. Some molecules are actually easier to study in distant parts of the Galaxy than in a laboratory. The ionized molecule N_2H^+ is so reactive that prior to its discovery in space it had never been possible to isolate enough of it for its structure to be studied in a laboratory. In the low density of regions of space, however, where its chances of being destroyed are reduced, the molecule exists in large quantities and produces clear spectral lines at wavelengths of a few millimeters. Scientists found they could measure finer details of the structure of the N_2H^+ molecule from signals that had travelled 30 000 light years across the Galaxy than from data they could produce in their own laboratories.

The third major factor that characterizes interstellar chemistry is low temperature. Gas atoms and molecules have very little kinetic energy. As a result, **endothermic reactions** – those which require the input of chemical energy – are practically irrelevant except in special circumstances like the heated gas behind an interstellar shock wave.

The fourth factor which makes interstellar chemistry different from terrestrial chemistry is the differing role of oxygen. Because oxygen is a highly reactive element, and because there is so much of it in the Earth's atmosphere, oxidation reactions, such as burning, affect our lives in major ways. In space, where oxygen atoms are greatly outnumbered by hydrogen, oxidation reactions are not nearly as important. This is one of the reasons that carbon monoxide is more common in space than carbon dioxide.

What kinds of reactions are left after all these limitations have been considered? Evidence suggests that under interstellar conditions the most important reactions are those in which one of the reacting particles is charged. These are referred to as **ion-molecule** reactions. They work well because the electric charge sets up an attractive force that pulls the reacting particles towards each other. Ion-molecule reactions involve only gas atoms and molecules, and do not make use of the dust grains. They depend on the fact that even in the darkest molecular cloud there are always a few ions. The number is small – something like one ion for every 10^8 molecules. They exist because molecular clouds, like everywhere else in space, are permeated by fast-moving protons called **cosmic rays** (see Chapter 9) that can strip electrons off atoms or molecules as they rush by. Chemical theories in which ion-molecule reactions play a dominant role have been successful in explaining some of the interesting features of Table 8.1, such as the presence of some obscure but highly reactive ionized molecules like HCO^+.

The fact that gas-phase ion-molecule reactions can explain most observations of molecular clouds is a relief to chemists; the main alternative theory, namely that most interstellar molecules are first formed on the surfaces of dust grains and then evaporate into space, is daunting in its complexity. The basis of such a theory is that the grain acts as a catalyst by facilitating contact among reacting molecules. The difficulty about such a theory is that each different kind of dust grain will affect each possible chemical reaction differently. Given our lack of knowledge about dust grain

88 MOLECULAR CLOUDS

surfaces it would be practically impossible to pin down what is happening if we had to worry about catalysts for every reaction.

Unfortunately, however, we cannot ignore grain-surface chemistry entirely. The most fundamental chemical reaction of all, namely the combination of two hydrogen atoms to form a hydrogen molecule, cannot be explained without recourse to dust grains. The difficulty is that when the two hydrogen atoms combine they have an energy surplus of up to 4.5 eV which must be accounted for somehow. Because of special quantum mechanical rules that apply to molecules made up of identical atoms, the newly-formed hydrogen molecule cannot simply radiate away a photon of the appropriate energy. Nor can it use the spare energy to rush off at high speed, since in a gas it has nothing to push against. Catalysis on grain surfaces appears to be the only way to account for the vast amount of molecular hydrogen found in our Galaxy. Reactions involving graphite grains may also be important for the production of the PAH molecules, whose link with dust grains we explored in Chapter 7.

8.5 Dense molecular clouds

Dense molecular clouds have been known in a different guise for decades. The distinctive patches of obscuration first noted by Herschel used to be termed **dust clouds** or **dark nebulae** (e.g. Figure 1.4). The gas content of these dust clouds was a mystery until the 1970s when it was realized that, in terms of mass, they contain far more gas, mainly in the form of H_2, than dust. We now usually refer to them as **dense molecular clouds** in recognition of their main ingredient – molecular hydrogen. The word

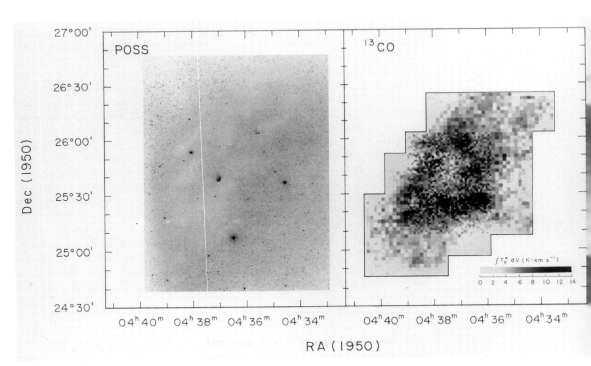

Figure 8.5. Molecular gas in dark clouds. On the left is a negative image (sky white, stars black) of a dark cloud in the Taurus constellation. On the right is a map of the same region made with a millimeter-wave telescope showing the emission from one of the isotopes of carbon monoxide. The thickest parts of the dark cloud show the greatest concentration of molecules.

'dense' is used to distinguish them from the transparent molecular clouds which are thin enough to be studied at ultraviolet wavelengths.

At about the same time as the nature of 'dust clouds' was being revealed, surveys of the sky to look for the 2.6-mm emission line of carbon monoxide led to the discovery of hundreds of giant molecular clouds in the disk of the Galaxy. Each of these giant clouds contains far more gas than anyone had ever dreamed could exist in an interstellar cloud. It was very soon realized that many giant molecular clouds lie right alongside prominent H^+ regions. Among the best examples is the Orion Nebula (Figure 5.3) which lies in front of one of the large Orion Molecular Clouds. The slightly different Doppler shifts of the gases in the H^+ region and in the molecular cloud show that they are separate entities which are in physical contact with each other. The nebula is eating its way into the molecular cloud. It is fortuitous that the H^+ region is on 'our' side of the molecular cloud; if the positions of the H^+ region and molecular cloud had been reversed we would have lost one of the finest sights in the sky.

Once a molecular cloud has been found it can be mapped by scanning across it with a radio telescope fitted with a receiver tuned to a transition of the CO molecule. Since many of these clouds are too far from the Earth to be visible by the obscuration they produce, the CO data provide the primary information on their shapes and sizes. We find an enormous variety of cloud features. The largest recognizable objects have diameters of about 80 pc, and sprawl across the sky. They seem to have fairly sharp edges, although we can guess that there is probably a layer round the outside of each cloud where the gas is partly molecular and partly atomic. Within each molecular cloud many smaller clouds are seen, with sizes ranging down to fractions of a parsec. The dense patches of the clouds are the parts most often likely to be associated with an H^+ region or an obscured source of infrared radiation. As we shall discuss in Chapter 12, they are also the most likely source of new stars.

A map of the CO emission provides a fine impression of a cloud's shape and size, but to answer more fundamental questions, such as whether or not the cloud is likely to spawn new stars, we need much more detailed information. How, then, can we use millimeter spectroscopy to determine things like the temperature, density, ionization and mass in the cloud? Because molecules are usually excited out of their ground states by collisions, the strength of a particular emission line usually depends on the temperature and density in the cloud. Different transitions depend on these quantities in different ways, so that the ratio of strengths of two lines can give us clues about the conditions in the cloud. By comparing two transitions of the same molecule we avoid problems arising from variations in a cloud's chemical make-up. The variety of different molecules found in molecular clouds can be turned to our advantage here, since different molecules are best suited for different jobs. The ratio of the strengths of certain NH_3 emission lines is a good indicator of gas temperature, while particle density can be estimated by comparing different transitions of the CS or CO molecules.

What these studies have shown is that throughout most of its volume the gas temperature in a molecular cloud is less than 15 K. The gas is cold mainly because the dust in the cloud manages to exclude starlight so well. Near the edges of the clouds, and in the vicinity of buried stars the temperatures can rise to 50 K or more. When this happens the clouds become powerful sources of infrared emission, and show up prominently on maps of the galactic plane produced by the IRAS satellite. The hottest regions are generally found at the interfaces between molecular clouds and H^+ regions, or in the immediate vicinity of newly formed stars. Gas densities are typically between 100 and 1000 hydrogen molecules cm^{-3} rising to

Figure 8.6. Photograph of the Orion constellation, one of the easiest to recognize in the sky. The Orion region includes many OB stars. The Orion Nebula (Figure 5.3) is visible as a fuzzy patch in the bottom center of the picture.

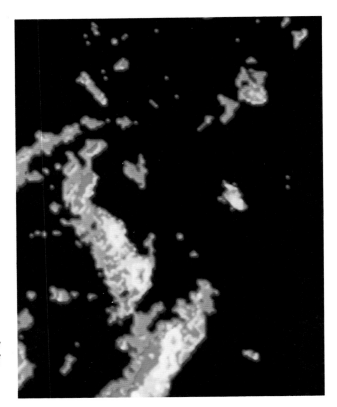

Figure 8.7. Molecular clouds in the same region of sky as Figure 8.7. These clouds have been mapped by measuring the strength of a spectrum line of the CO molecule at a wavelength of 2.6 mm.

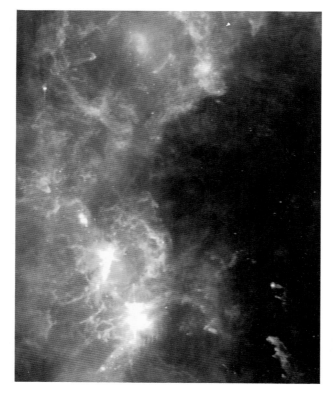

Figure 8.8. Infrared emission from the same region of Orion as Figure 8.7. The hottest dust coincides with the dense molecular clouds. The dust in the molecular clouds is heated by a combination of internal and external stars. Note also the circular ring of warm dust around the star λ Orionis at the top of the picture.

more than 1 000 000 in the dense cores of clouds where new stars are forming. The total mass of a giant molecular cloud complex can reach 10 000 000 times the mass of the Sun, making it the most massive type of object in our Galaxy.

Astronomers face some serious problems in trying to determine the properties of a molecular cloud. One is the nagging feeling that concentrating attention on a molecule which is diluted by a factor of 10^4 may be akin to trying to reconstruct a dinosaur from a single fossilized tooth. A more immediate problem is that with some molecules we can only see the front of the cloud. In a giant molecular cloud the emission from carbon monoxide is so strong that it is **self-absorbed**; so many photons are produced that they become prey to stimulated absorptions by other carbon monoxide molecules. When this happens photons from the rear portions of the cloud are absorbed before they reach us, and our data on the cloud become incomplete and difficult to interpret. What makes the problem particularly difficult to deal with is the fact that the amount of self absorption can be different at different Doppler velocities within the emission line; the correction for self absorption depends on how the gas in the cloud is moving, as well as on how dense it is. An interesting way around this problem is to observe the emission from carbon monoxide molecules in which one of the atoms is a rare isotope. One carbon atom in 90 has an extra neutron in it, making its atomic weight 13 instead of 12. The molecule it forms is referred to as ^{13}CO. One oxygen atom in 500 has two extra neutrons and an atomic weight of 18 instead of 16. The molecule it forms is called $C^{18}O$. The three variants of the carbon monoxide molecule produce emission lines of slightly different wavelength which can be distinguished by a radio telescope. Self absorption is much less of a problem for the rare isotope lines because there

92 MOLECULAR CLOUDS

are so many fewer molecules able to cause interference. For this reason they are often used as probes of the dense cores of molecular clouds.

8.6 Tools of the trade: millimeter-wave astronomy

Covering the wavelength range from about 0.3 to 3 mm, millimeter-wave astronomy grew out of radio astronomy during the 1970s and 1980s, but borrowed a number of techniques and problems from infrared astronomy. The impetus for the exploitation of this waveband was the discovery of the 2.6-mm emission line of carbon monoxide, plus a large number of other emission lines with comparable wavelengths. To study these lines astronomers use a millimeter-wave telescope attached to a radio spectrometer – a bank of receivers all tuned to slightly different wavelengths.

Millimeter-wave telescopes look much like radio telescopes, but because the waves are so much more compact their reflecting surfaces have to be much more precise. It is hard to make very large parabolic dishes with sufficient accuracy, so millimeter-wave telescopes are usually smaller than radio telescopes. The largest is currently the 45-m diameter reflector at Nobeyama in Japan, but most have diameters of 20 m or less. Some telescopes, particularly those designed to operate at the shortest wavelengths, are housed in a radome or a building which can be closed during bad weather.

Figure 8.9. Two of the millimeter-wave antennas of the Institute Radioastronomie Millimetrique (IRAM) near Grenoble, France. Each antenna is 20 m in diameter.

Besides the mechanical problems of constructing an accurate reflector and making sure it can point to precisely the right place in the sky, the main difficulties faced by millimeter-wave astronomers are designing sensitive receivers and worrying about the weather. At around 2–3 mm wavelength radio telescopes can operate reasonably well from many sites around the Earth, but at submillimeter wavelengths (i.e. shorter than 1 mm) absorption by water vapor in the Earth's atmosphere starts to become significant. Telescopes designed for these wavelengths must be sited at places where the atmosphere is particularly dry. The two leading submillimeter telescopes, the 10-m diameter Caltech Submillimeter Observatory, and the 15-m UK–Canada–Netherlands James Clerk Maxwell Telescope, are both sited on the 4000-m summit of Mauna Kea in Hawaii. In the USA, major millimeter-wave telescopes are also sited in Arizona, California and Massachusetts, while consortia of European astronomers operate facilities in Spain, France, and Chile.

Although the major motivation for building millimeter-wave telescopes remains the study of interstellar molecules in our Galaxy and in other galaxies, they are useful for other studies as well. Interstellar dust grains with temperatures of 10–40 K are strong sources of power at wavelengths shortward of 1 mm. The emission from grains is thermal radiation, so they emit a continuous spectrum rather than emission lines. Another use for millimeter-wave telescopes is studying the variations in the power output of quasars.

8.7 Living in a molecular cloud

Would life be any different if the Sun were inside a molecular cloud instead of at its current position in the Galaxy? The extinction inside one of these clouds is so large that we would see nothing beyond the edge of the Solar System. Although the Sun, the planets, and their moons would look much as they do now, there would be no background of stars, and no galaxies and nebulae to view through telescopes. The only compensation is that we might see more comets than we do now, but on some moonless nights there would be nothing to see in the sky at all. For those of us with an interest in astronomy the loss in information would be catastrophic. Everything we know about the Universe that is based on visible, ultraviolet and X-ray observations would be lost, although if we had infrared and radio telescopes we would be able to use them to see out of the cloud.

It is not only astronomers and romantics whose lives who would be affected by the loss of the starlit sky. The history of the last few centuries might have been very different. For one thing, the lack of fixed navigation points would have probably set back the exploration of the oceans by hundreds or even thousands of years. The patterns of international settlement, and of naval power, would have been totally different if mariners had not had the confidence to sail far from known coastlines. It is also extremely doubtful whether our current technological civilization could have risen at all if the stars had not been there. A turning point in the history of science was the discovery by Johannes Kepler that the planets move in elliptical paths. This work led directly to Isaac Newton's formulation of the laws of gravitation and motion, and triggered a revolution in science that is still going on to this day. Kepler would almost certainly not have come up with the theory of elliptical orbits if the stars had been invisible. The unmoving distant stars provided him with fixed reference points against which the motions of the planets could be logged. Without them, he would have been faced with a much harder problem of disentangling the motions of the planets from those caused by the motion of the Earth as it spins in its orbit around the Sun.

The absence of starlight would have affected the history of life on Earth even before human intelligence developed. For one thing, many mammals are nocturnal; their ability to find food, and hence evolve, would have been impaired on moonless nights; they might have developed infrared-sensitive 'eyes' like some snakes. For another, we know that migrating birds use the stars as one of their ways of navigating over long distances. Seeds are spread round the world in the feathers and stomachs of birds, so if migration had evolved in a different way, the pattern of the world's flora would have been affected just as much as its fauna.

8.8 Molecular clouds in the Galaxy

There are several thousand giant molecular clouds like those in Orion. They can be detected at great distances, and surveys at the wavelength of the 2.6-mm CO line have become an important tool for studying the structure of the Galaxy. The surveys are made by recording the shape of the 2.6-mm spectral line in an array of positions around the sky. As with the 21-cm data we discussed in Chapter 4, the distance to a cloud is estimated from its Doppler shift and from our knowledge of how the Galaxy rotates.

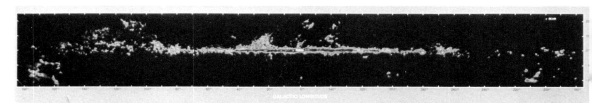

Figure 8.10. Molecular clouds in the Galaxy. Like the H^+ regions, the molecular clouds are generally seen very close to the galactic plane.

The total mass of molecular hydrogen within a radius of 14 kpc of the center of the Galaxy is around 2×10^9 solar masses, which is about the same as the total mass of atomic hydrogen, as determined from the 21-cm line. The ways that the two kinds of gas are distributed around the Galaxy are very different, however. The majority of molecular clouds are found either within 2 kpc of the Galaxy's center, or in a broad ring that circles it between radii of 3 and 7 kpc. There are comparatively few molecular clouds in the outer reaches of the Galaxy. In contrast, atomic hydrogen is scarce near the galactic center, but quite evenly distributed between radii of 3 and 14 kpc. Overall, we find that in terms of mass (but not volume), molecules are the dominant form of interstellar gas in the inner parts of the Galaxy, while atomic gas prevails towards its edge.

The H^+ regions in our Galaxy (Chapter 5) are distributed more like the molecular clouds than the atomic gas, in that they are concentrated in the 3–7 kpc distance range. The reason for this correlation is the stars that ionize H^+ regions live such short lives that they never stray far from the place of their birth in a molecular cloud.

Figure 8.11. Molecular, atomic and ionized gas are distributed very differently in the Galaxy. Molecular and ionized gas is concentrated in a ring of clouds and H^+ regions between about 3 and 7 kpc towards the inner Galaxy, while atomic gas is found in abundance at much larger distance from the center. There is also a very strong concentration of molecular gas within 2 kpc of the galactic center.

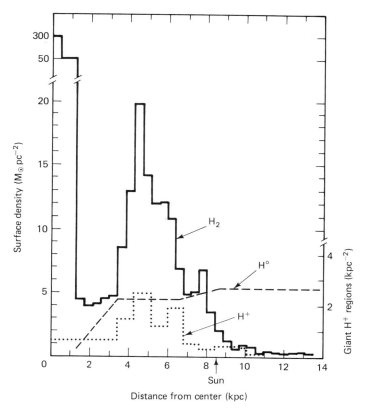

8.9 Interstellar masers

The first interstellar molecule to be discovered at radio wavelengths, the hydroxyl (OH) molecule, is still the most difficult to understand. Deep within certain molecular clouds there are pockets of gas which radiate OH emission lines that are millions of times more intense than can be accounted for by straight-forward collisional excitation. The emission lines come from very small volumes; a typical OH source displays emission from a dozen or more 'hot-spots', each smaller than the Solar System. The spots have slightly different Doppler shifts and usually vary in intensity over a period of months.

These effects can be explained only if the OH molecules are acting as an interstellar **maser** – an acronym for **m**icrowave **a**mplification by **s**timulated **e**mission of **r**adiation. Within the maser region an anomalously large number of OH molecules are somehow forced out of the ground state into one of the upper energy levels. It is a consequence of quantum mechanics that if a photon of just the right wavelength travels through this region it can trigger downward transitions of molecules out of the upper state. This process is called **stimulated emission**. It is exactly analogous to the much more common process of stimulated absorption, except that photons are created instead of absorbed. Since the new photons that are generated have the same wavelength as the photon that triggered the emission, the overall effect is that the intensity of the original radiation is amplified by its passage through the maser region. Because the amplification process is cumulative, it can generate intense beams of radiation very efficiently. The same principle is the basis of the **laser**, except that the laser uses light instead of microwaves. The difficulty for astronomers comes in trying to explain how to get enough OH molecules into one particular excited state. Collisional excitations are not enough; some other processes such as radiative excitations or chemical reactions must be invoked as well. Theoretical calculations show that this masering process can take place only under very special conditions of density, temperature, gas motions and magnetic field, although astronomers have not yet agreed on what these conditions are.

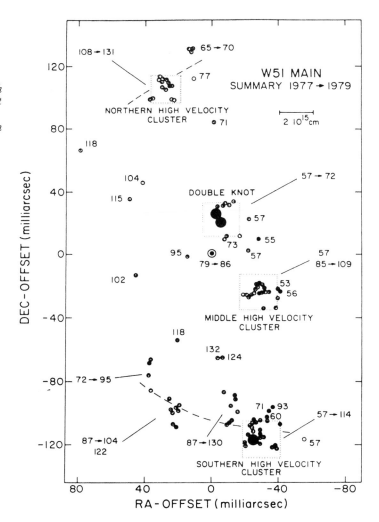

Figure 8.12. Maser emission from H_2O molecules in the W51 cloud. Maser emission is concentrated into intense spots within a molecular cloud. The numbers near each spot refer to the Doppler shift in km s^{-1} of the emission at that point. No visible light can be detected from this cloud.

The original OH masers were discovered at a wavelength of 18 cm. Subsequently, maser emission from OH molecules has been found at other, shorter, wavelengths. Several other molecules are now known to exhibit maser activity as well as OH; the most notable of these are water vapor (H_2O), and silicon monoxide (SiO). Water vapor masers are even more powerful than OH masers; the source called W49, for example, emits as much power in a single spectral line at 1.35 cm as does the Sun at all wavelengths. Hydroxyl and water masers are often found close to each other in the same molecular cloud, but they are not actually coincident.

Masers are found either in the immediate surroundings of giant stars, or in the cores of molecular clouds. Because only tiny regions of a molecular cloud ever act as masers they do not provide us with much direct information about the interstellar medium. The fact that OH and H_2O masers are so bright, however, makes them useful in some indirect ways. First, they can act as beacons that draw astronomers' attention to the hottest and densest regions of molecular clouds. These regions are of particular interest because they may be the birthsites of newly-formed stars. Many of the early infrared studies of molecular clouds focused on the immediate vicinity of masers. Second, H_2O masers are so bright and so small that changes in their positions can be measured to enormous accuracy. By combining the signals from radio telescopes separated by thousands of kilometers astronomers have been able to watch dense knots of gas being expelled bullet-like from certain newly-formed stars.

9 Cosmic rays and magnetic fields

There is one group of interstellar particles that does not fit into any of the categories we have so far described. These are the **cosmic rays**, a mixture of electrons, protons, and atomic nuclei which rush around the Galaxy at speeds close to the speed of light. The somewhat confusing use of the word 'ray' to describe a particle dates from a time when their nature was still a mystery. Cosmic rays are found throughout interstellar space, in regions that are both hot and cold, dense and diffuse, ionized and neutral. While cosmic rays interact with ordinary interstellar matter in some important ways, they behave for the most part as an independent fluid, almost unaffected by the forces which mould the interstellar clouds through which they pass.

Only one interstellar particle in a billion is a cosmic ray, but because they have so much kinetic energy, cosmic rays have an importance to the interstellar medium out of all proportion to their abundance. It is important to keep in mind that cosmic ray particles are a special breed, and not simply interstellar protons that happen to have higher than average speeds. The velocity of a typical cosmic ray proton is at least *100 000 times* greater than that of a typical interstellar atom, and the kinetic energy it possesses is enough to ionize at least *10 000 000* hydrogen atoms.

Unlike other types of interstellar matter, cosmic rays *can* be studied directly. Their high speeds carry them right through the Solar System to the vicinity of the Earth. Screening by the Earth's atmosphere stops most cosmic rays reaching sea level, but they can be collected at the tops of high mountains, or from rockets, balloons, or satellites. We can find out what they are made from and how fast they are moving. What we *cannot* tell is where they are coming from. Because cosmic rays are electrically charged they are subject to electric and magnetic forces that do not affect photons or neutral particles. The paths of cosmic rays are continuously deflected in different directions as they encounter the constantly-changing magnetic fields of interstellar space. The direction in which a cosmic ray is moving when we intercept it therefore tells us almost nothing about its past history. The observational problems faced by cosmic ray astronomers are analogous to those that optical astronomers would face if they were forced always to observe the skies through a frosted glass window which randomly scattered all starlight as it entered the telescope.

The link between cosmic rays and magnetic fields is so strong that we must start this chapter with a general discussion of the way that charged particles and magnetic fields affect each other.

9.1 Fields and particles

The human body is not sensitive to magnetic fields. However, we can appreciate their existence when we see a compass needle deflected by the Earth's magnetic field, or when we see magnetized objects attracted to metal surfaces like refrigerator doors. What is more difficult to visualize are the fundamental links between magnetism and electricity; moving electric charges generate magnetic fields, while magnetic fields exert forces on moving charges. These ideas are central to the technology of electric motors and generators. Inside these machines moving electric charges are harnessed in the form of currents in copper wires.

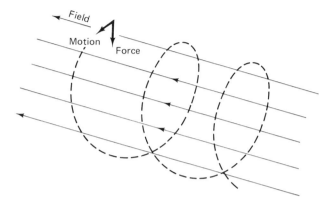

Figure 9.1. The force on a proton in a magnetic field is at right angles to the direction of its motion. The path that the particle follows is generally a helix.

It is useful to visualize a magnetic field as a series of parallel lines forming a kind of stream flowing in a particular direction. In a laboratory the magnetic field lines might connect the poles of a piece of magnetized metal, but in space, where permanent magnets are virtually absent, the streams always join back on themselves as loops of some shape or another. The force that a magnetic field exerts on a charged particle depends on its motion relative to the field. A charged particle that is stationary or that is moving along the direction of the field feels no force, and is not deflected. A charged particle moving across the field, on the other hand, feels a force that is at right angles both to the magnetic field and to the particle's direction. The particle is therefore deflected into an arc that curves it around the magnetic field. If the magnetic field is uniform over a large enough volume the particle will execute repeated circles around the magnetic field lines; the faster the particle and the weaker the magnetic field, the larger the radius of the circle. If the particle moves at an angle to the field then its subsequent motion will have the shape of a helix; if the magnetic field is not uniform the shape of the path will be more complicated. A cosmic ray travelling through the Galaxy follows an extremely tangled path.

When a charged particle moves through a region of space containing a steady magnetic field the direction of its motion changes, but not its energy. In other words the net speed of the particle stays constant. Only magnetic fields which are changing with time can increase the energy of a charged particle.

9.2 Cosmic rays in the Galaxy

Cosmic rays travel with speeds close to that of light; because of this, their motions follow the laws of Einstein's theory of **relativity** rather than those of the much simpler Newtonian physics. Fundamental to the theory of relativity is the idea that nothing can travel faster that the speed of light. As a particle approaches this speed it becomes more and more difficult to accelerate, an effect that is equivalent to saying that its mass increases. A particle traveling in this regime is said to be **relativistic**, and its motion is then better described in terms of its kinetic energy rather than by its speed, since the latter is so close to the speed of light. Protons with a kinetic energy of 10^9 eV, for example, travel at 88% of the speed of light. A proton with ten times greater kinetic energy moves only about 13% faster – at 99.6% of the speed of light – but its effective mass is five times larger.

The bulk of the cosmic ray particles in the vicinity of the Earth have energies around 10^8–10^9 eV, but many particles with much greater energies are also seen. Figure 9.2 shows how the number of protons of a given energy depends on that energy. Above 10^9 eV the number of particles drops steadily with increasing energy all the way to 10^{20} eV. We know very little about cosmic rays with energies greater than 10^{19} eV because so few of them arrive on the Earth each day. There is, nevertheless, great interest in such particles because they have energies vastly greater than can be generated by particle accelerators in nuclear physics laboratories. The energy contained in a single 10^{20} eV cosmic ray proton is 16 joules, enough to lift this book several centimeters off the table. The speed of such a particle differs from the speed of light by less than a millimeter a century.

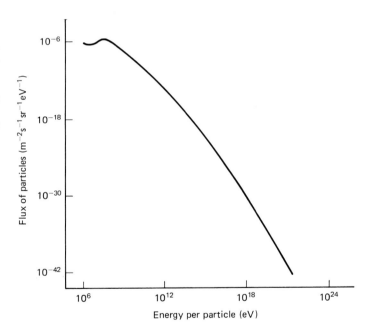

Figure 9.2. The number of cosmic rays as a function of their energy. At low energy the cosmic rays, which are mainly protons, are detected by satellite- or balloon-borne experiments. At energies above 10^{14} eV information comes from arrays of detectors on the ground.

Energetic cosmic rays possess so much momentum that they are essentially undeflected as they pass through the Solar System towards the Earth. Cosmic rays with energies less than 10^8 eV, however, are much more susceptible to the influence of the magnetic field of the Sun and of the ever changing **solar wind** of ions that stream out from it. As a result, comparatively few particles with energies less than 10^8 eV reach the Earth, and our knowledge of the flux of low-energy cosmic rays in the Galaxy is severely limited.

From an analysis of the masses and charges of cosmic rays at the top of the Earth's atmosphere we know that some 90% of the particles are protons,

1% are electrons, and that most of the rest are nuclei of helium atoms. In general, the pattern of element abundances in cosmic rays is similar to that of the Solar System, but there are some important differences. Lithium, beryllium and boron are at least 10 000 times more abundant in cosmic rays than in the Solar System or in interstellar space. These extra atoms are produced by nuclear reactions that occur as the result of collisions between fast-moving cosmic ray particles and stationary atoms of interstellar matter. It has been calculated that a typical cosmic ray would have to travel a distance of 10^6 parsec through the Galaxy's disk in order to produce the observed amounts of lithium, beryllium and boron found in cosmic rays. The fact that this distance is 100 times greater than the diameter of the Milky Way is an indication of the tortuousness of the path that a cosmic ray takes as it traverses the Galaxy.

The only place in the Galaxy where we can measure cosmic rays directly is in the vicinity of the Solar System. Cosmic rays in other parts of the Galaxy can be indirectly detected in two ways, however. We can collect γ-rays produced by collisions between cosmic rays and interstellar matter and we can map the radio-wavelength synchrotron emission produced by cosmic ray electrons as they spiral around the galactic magnetic field.

9.3 Gamma rays from the Galaxy

Gamma rays are the highest energy, and hence shortest wavelength, photons that are observed by astronomers. Because they are photons rather than charged particles they pass undeflected through the galactic magnetic field. Figure 9.3 is a map of the sky as it appears for γ-rays in the wavelength range 70–5000 MeV. Most γ-rays come from a band of emission that follows the plane of the Milky Way, although a minority originate from discrete objects such as the Crab and Vela supernova remnants (see Chapter 10).

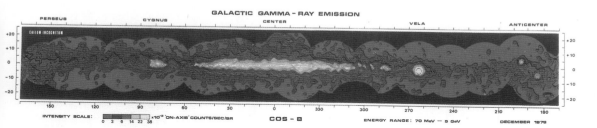

Figure 9.3. Gamma rays from the plane of the Galaxy. The yellow areas are the regions of highest intensity. This map was produced from data collected by the COS-B satellite.

There are two important ways by which cosmic rays can generate γ-rays. The first is by **pion decay**. When relativistic cosmic ray protons collide with hydrogen atoms in the interstellar medium, nuclear reactions take place between them that result in the formation of subatomic particles called pions. These particles are unstable and decay almost immediately to produce a pair of γ-rays. These γ-rays have an average energy of about 100 MeV each, corresponding to a wavelength of 10^{-14} m, and can pass right through the Galaxy with almost no hindrance by the interstellar medium.

The second way of generating γ-rays is by the **bremsstrahlung** or 'braking-radiation' process. It is a fundamental property of an electrically charged particle that whenever its velocity is changed, electromagnetic radiation is generated. One good way that a charged particle can be accelerated is by attracting it towards a particle which has opposite electrical charge. This process happens in the interstellar medium whenever a fast-moving electron passes close to the nucleus of an interstellar atom. As the electron rushes through the interstellar medium it is tugged repeatedly from side to side, generating electromagnetic radiation at each impulse. As it does so it slows down, its kinetic energy being converted into γ-rays which we observe. Cosmic ray protons also emit bremsstrahlung radiation as they pass through interstellar matter, but because they are much heavier than electrons, their accelerations are much smaller, and the radiation they produce is insignificant.

The pion decay and bremsstrahlung processes both involve interactions between cosmic rays and the *nuclei* of atoms. Because the kinetic energy of a cosmic ray is so much greater than the ionization potential of an atom, it makes no difference whether the interstellar nucleus is an isolated ion, or part of a neutral atom, a molecule, or even a dust grain. When cosmic rays permeate interstellar clouds the amount of γ-radiation they produce depends only on the number of cosmic rays and on the density of the interstellar matter. Figure 9.3 is therefore effectively a map of how the *product* of the interstellar density and the cosmic ray density varies in different parts of the Milky Way. What it shows is that this product is heavily concentrated towards the central regions of the Galaxy in a way that is highly reminiscent of the distribution of the galactic molecular clouds as revealed by millimeter-wave observations of the CO molecule (Figure 8.10). The strong similarity between the γ-ray and CO maps leads to three conclusions. First, it demonstrates that cosmic rays pervade the whole Galaxy, not just the neighborhood of the Sun. Second, it confirms the statement made in Chapter 8 that in the inner regions of the Galaxy the molecular clouds are a dominant form of interstellar matter. Third, it has enabled astronomers to calibrate the relationship between CO line intensity and molecular hydrogen column density in a way that does not involve assumptions about the properties of interstellar dust.

What does Figure 9.3 tell us about the distribution of cosmic rays in the Galaxy beyond the fact that they are widespread? To answer this question it is helpful to consider the energies of the γ-rays being collected; the higher energy photons come mainly from the pion decay resulting from *proton* collisions, while the lower energy ones derive more from *electron* bremsstrahlung. When the γ-ray maps are compared in detail with maps of the interstellar H^0 and H_2 density there are indications that the intensity of cosmic ray electrons is slightly greater towards the center of the Galaxy that towards its outer regions. The corresponding variation in the intensity of cosmic ray protons is almost zero. Certainly, the intensity of either type of cosmic ray changes much less with galactic position than does the density of either stars or interstellar matter. Whatever their source, cosmic rays are very efficient at distributing themselves rather evenly around the Galaxy.

9.4 Tools of the trade: cosmic ray and gamma ray telescopes

Cosmic ray detectors bear very little resemblance to astronomical telescopes. They have neither lenses nor mirrors, and make no attempt to form any kind of image of the sky; since the paths of cosmic rays are so convoluted, the direction of travel of a cosmic ray is almost irrelevant anyway. The interesting things that can be measured about cosmic rays are their kinetic energies, their masses, and their charges. By measuring large numbers of cosmic rays a statistical picture can be built up describing the relative numbers of nuclear types and, within each type, the relative numbers with different kinetic energies.

There are different kinds of cosmic ray detector, but most make use of the fact that a cosmic ray possesses so much energy that it leaves a substantial trail when it passes through matter. One common device, the spark chamber, consists essentially of a gas-filled box with a high voltage across it. When a cosmic ray passes through the chamber it collides with many of the gas atoms and leaves behind it a trail of ions and electrons. These charged particles are collected by electrodes near the edges of the box and produce a measurable pulse of electric current. The principle is similar to that in a Geiger counter. Photographic film is also used for detecting cosmic rays, but no light is involved. Cosmic rays that pass through the film leave tiny marks in the photographic emulsion that may be measured and counted when the film is developed.

Most cosmic rays are stopped by the Earth's atmosphere, so experiments to study them have to be carried aloft in satellites or under high-altitude balloons. Very high energy cosmic rays, however, can be studied at ground level via the effects they produce when they strike the Earth's upper atmosphere.

Figure 9.4. Cutaway picture of NASA's Gamma Ray Observatory, launched in 1991. This satellite contains four different γ-ray telescopes, capable of detecting photons with energies between 20 keV and 30 GeV. The telescope allows astronomers to determine the positions of γ-ray sources to within a few arcminutes.

Although γ-rays are photons, the instruments used to study them are much more like cosmic ray detectors than optical or radio telescopes. The problem with γ-rays is that they can pass though substantial thickness of metal or glass, so that they cannot be focused by mirrors or lenses. Gamma ray telescopes contain instruments that register the arrival of each individual photon, and calculate the approximate direction from which it came. The COS-B satellite, which was launched in 1975 and which produced the map of the Milky way shown in Figure 9.3, had an angular resolution of only about 2°, giving us a view of the Universe that is much more blurred than what we get at radio or visible wavelengths. Another problem is that there simply are not very many γ-ray photons to be observed; typical collection rates for COS-B were a few photons an hour! For comparison, the unaided human eye needs to receive many hundreds of photons a second to register even the faintest visible star.

9.5 Galactic radio emission

The helical path of a cosmic ray around the galactic magnetic field is an accelerated motion, and it can therefore generate electromagnetic waves. This emission is called **synchrotron radiation**. Electrons, being much lighter and more easily accelerated, produce much more synchrotron emission than protons and under typical interstellar conditions they radiate most strongly at radio wavelengths. The synchrotron process, in fact, is by far the most widespread source of astronomical radio emission, both from within our Galaxy and beyond it. The diffuse emission from the Galaxy is most powerful at the longest wavelengths that can be observed from the Earth — a few meters — and declines steadily with decreasing wavelength until it becomes difficult to disentangle from other sources of radio noise at wavelengths of a few centimeters or less.

A map of the diffuse radio emission from the Galaxy is shown in Figure 9.5. There is a concentration towards the galactic plane, but there are other prominent features as well. The largest of these, the North Polar Spur, rises from the plane near galactic longitude 30°, and may be the result of a supernova explosion some 10^5 years ago (see Chapter 10). By comparing

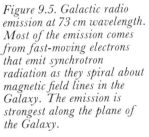

Figure 9.5. Galactic radio emission at 73 cm wavelength. Most of the emission comes from fast-moving electrons that emit synchrotron radiation as they spiral about magnetic field lines in the Galaxy. The emission is strongest along the plane of the Galaxy.

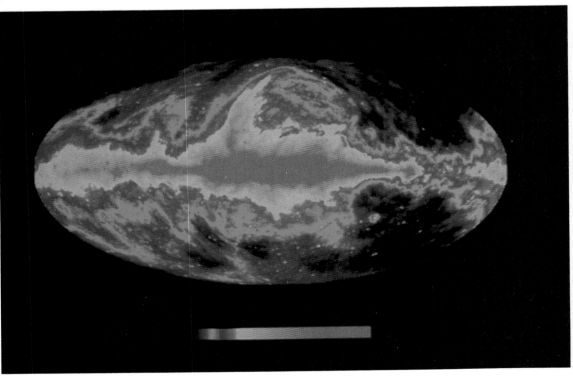

Figure 9.5 with Figure 8.10 it can be seen that the thickness of the galactic plane is greater at radio wavelengths than at millimeter wavelengths. The implication of this difference is that the cosmic ray electrons are less confined to the plane of the Galaxy than are the molecular clouds. Figure 9.5 does not tell the whole story about radio emission from the Galaxy. When the galactic plane is observed in detail it is seen to be filled with large numbers of discrete radio sources superimposed on the broad background shown in the picture. These sources, which have typical angular diameters of a few arcminutes each, are mainly of two kinds. There are H^+ regions (Chapter 5) associated with young, hot stars, in which the radio emission is free–free emission caused by collisions between electrons and ions moving at speeds of a few kilometers a second, and there are supernova remnants (Chapter 10) in which the emission is synchrotron radiation from relativistic electrons. Both emission mechanisms produce continuum emission, but they can be distinguished by the fact that free–free emission becomes stronger at short wavelengths, while synchrotron emission becomes weaker. Synchrotron emission may also be distinguished from free–free emission by the fact that it is usually slightly polarized (see Section 9.8).

9.6 Sources of cosmic rays

Astronomers are not at all sure where cosmic rays come from. Two questions need to be addressed. First, what is the origin of the particles themselves, and second, how did they become accelerated to relativistic speeds. The answers to these questions may or may not be related. One possibility is that cosmic rays are produced and simultaneously accelerated in supernova explosions. Supernovae are powerful enough, and, as demonstrated by their synchrotron radio emission, are capable of accelerating electrons to relativistic speeds. The relative abundances of the elements in cosmic rays, however, are not what we would expect in the material ejected in a supernova explosion. An idea that is currently gaining favor is that the cosmic ray particles start out as the winds from stars quite similar to the Sun, and are accelerated to relativistic speeds by magnetic fields associated with shock waves in the interstellar medium.

9.7 The nature of interstellar magnetism

Magnetic fields are sustained by electric currents. In interstellar space these currents arise from the systematic motions of large numbers of charged particles. Electrons do most of the charge transport, since they are lighter than ions. They play this role not only in H^+ regions where they are numerous, but also within atomic and molecular clouds, where a small, but significant percentage of atoms is kept in an ionized state as a result of impacts by cosmic rays. We cannot detect the motions of the charged particles directly, and have very little basis on which to estimate the paths of the currents. Even if we knew them they would not be very useful because the magnetic field at any point in space is made up of contributions from moving charges all around it; currents arising from the large scale rotation of the whole Galaxy can influence the magnetic field as much as local irregularities.

Although it is difficult to unravel how the Galaxy's magnetic field got started, it is comparatively easy to understand how it can be changed as a result of the motions of the interstellar gas. The key to this understanding is the notion of **flux freezing**. Consider an isolated interstellar cloud, as in Figure 9.6, which is threaded by a magnetic field which we denote by parallel lines; the closer together the lines, the stronger the magnetic field. Suppose now that the outer parts of the cloud start to contract towards its center – perhaps as the prelude to the formation of a star. The laws of

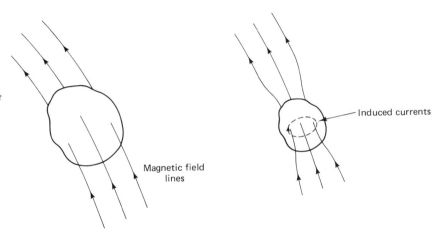

Figure 9.6. *Flux freezing in an interstellar cloud. If a cloud has a magnetic field threading it, and the cloud then collapses, currents are induced that make the field inside the cloud stronger. The lines of magnetic flux are effectively 'frozen' into the cloud.*

electromagnetism tell us that as the moving electrons in the cloud encounter the magnetic field lines they feel a sideways force that accelerates them into paths that circle the magnetic field lines. The circling motion of these moving electrons constitutes an electric current which generates a magnetic field that adds to the field inside the cloud and subtracts from the field outside it. The net result is that as the cloud collapses the lines of magnetic field move inward with the material. The electrons which have been induced to move within the cloud continue to do so indefinitely, maintaining the new stronger field until such time as the shape of the cloud changes again. The magnetic field is thus tightly linked to the motions of the electrons which are themselves tied to the rest of the gas by electrostatic attraction and by frictional forces. The magnetic field lines, sometimes known as the magnetic flux, are in effect 'frozen' into the interstellar gas via their interactions with the electrons.

There are some interesting consequences of flux freezing. For one thing, it means that gas can move comparatively easily along magnetic field lines, but only with difficulty along paths which cross them. This idea has been used to explain why some interstellar clouds have long thin shapes. For another, it means that the magnetic field of the Galaxy is greatly affected by the large-scale motions of the interstellar gas. Stars and gas at different distances from the center of the Galaxy revolve around it at different speeds. Interstellar clouds at slightly different radii therefore slowly become strung out around the galaxy as it rotates, a mechanism that also contributes to the formation of spiral arms. Any magnetic field associated with the clouds also gets stretched around the Galaxy, even if it started out by pointing in a direction directly toward the center (Figure 9.7). Flux freezing therefore causes the magnetic field of the Galaxy to become wound around like string on a reel. This winding-up process also boosts the strength of the magnetic field since the lines are forced more closely together as they are stretched by the differential rotation.

9.8 Measuring the galactic magnetic field

The magnetic field of the Galaxy is extremely hard to determine. There is no way to measure it directly, because within the Solar System it is swamped by the much stronger fields produced by the Earth and the Sun. There are several indirect methods that can be used, but unfortunately they all involve measurements of rather subtle effects and rarely produce data which can be interpreted unambiguously.

The longest-established method for studying the galactic magnetic field

Figure 9.7. Differential galactic rotation can amplify the magnetic field in the Galaxy and stretch it out into an approximately circular pattern, whatever its original direction.

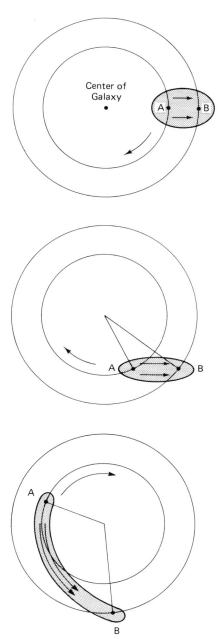

involves measuring the polarization of starlight that has been reddened by interstellar dust. As mentioned in Chapter 7 the polarization arises from the interplay of two physical effects that only work on grains that have certain shapes. When a long, thin grain is subjected to a magnetic field in an interstellar cloud it tends to line itself up so that its long axis is at right angles to the magnetic field; impacts from atoms in the cloud make the grain spin like a little propeller around the magnetic field line. If the magnetic field is in the same direction throughout the cloud all the long thin grains will be spinning around the same axis. When light from a star passes through a cloud in which many grains are magnetically aligned,

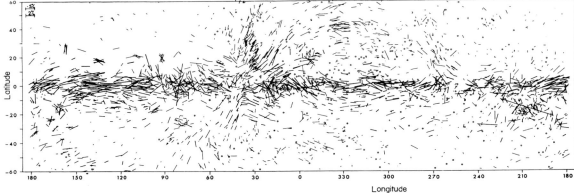

Figure 9.8. Polarization of starlight in and around the galactic plane. Each line represents one star. The length of the line is proportional to the amount of polarization while the direction of the line shows its angle. Longitude 0° points towards the center of the galaxy; longitude 180° points directly away from the center.

more radiation of one polarization is absorbed than the other. The direction of polarization of the starlight we receive on Earth indicates the direction of the magnetic field where the extinction took place. Figure 9.8 shows that there is some clear coherence in the patterns of the magnetic field around the sky. Stars at low latitudes are generally polarized parallel to the galactic plane except near longitudes 80° and 270° – approximately along the direction of the Solar System's orbit around the Galaxy. This is the pattern we would expect for a magnetic field that circles around the Galaxy, as predicted on the basis of flux freezing and differential rotation. Away from the galactic plane the picture is less simple; near galactic longitude 40° the magnetic field appears to be stretched in the same direction as some of the hydrogen gas seen in Figure 4.9, and the radio emission seen in Figure 9.5 – a strong indication that the shapes of these gas clouds have been affected by magnetic forces.

The data on optical polarization suffer from the problem that plagues all measurements of starlight: our view of the Galaxy is restricted by extinction to distances of about 1 kpc. To see patterns on a larger scale our best guide is to look at other galaxies which we believe to resemble the Milky Way. The way we do this is to make use of the fact that the synchrotron radio emission given off from galaxies is intrinsically polarized. The polarization has nothing to do with dust grains, but arises from the fact that the cosmic ray electrons which produce the synchrotron emission are confined to move in helical paths around the magnetic field of the galaxy. The pattern of polarization seen in the radio emission therefore reveals the magnetic field of the galaxy, at least in outline. Figure 9.9 shows that the galaxy NGC 6946 has a roughly circular pattern of magnetic field, similar to what has been deduced for the vicinity of the Sun in our Galaxy.

Within the Milky Way Galaxy synchrotron radio emission is a poor guide to the direction of the magnetic field. It can provide us with a useful estimate of typical field strength, however, since the amount of synchrotron radio emission given off by a relativistic electron depends strongly on the strength of the magnetic field it is spiralling round. If we assume that the cosmic ray electrons that we collect above the Earth's atmosphere are typical of those producing the galactic radio background we can deduce that the magnetic field in the Galaxy is about 3 microgauss. By human standards this is a very weak field indeed, nearly a million times less than that of the Earth.

The galactic magnetic field as derived from the strength of the synchrotron emission is only a rough average value. It tells us nothing about the differences in field strength in different phases of the interstellar medium. To study these variations we can make use of two further phenomena, namely **Faraday rotation** and the **Zeeman effect**. Faraday rotation

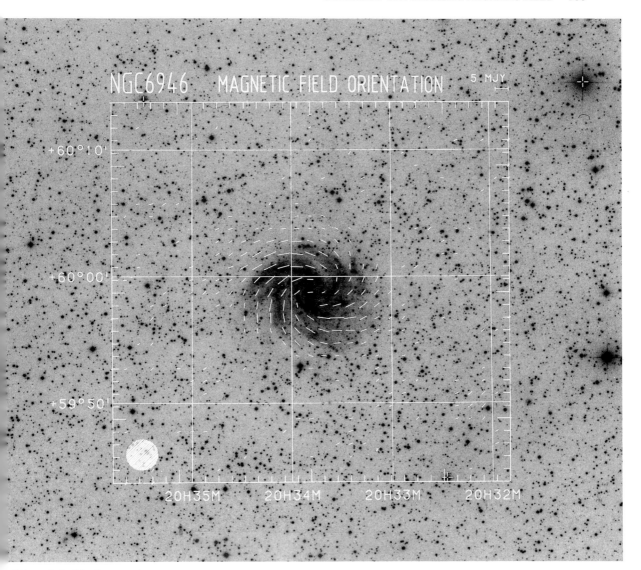

Figure 9.9. The large scale magnetic field in the galaxy NGC 6946. The direction of the magnetic field is deduced from the polarization of the synchrotron radio emission.

occurs when a polarized radio wave passes through a region of space which contains both ionized gas and a magnetic field. The angle of the wave's polarization rotates by an amount which depends on the density of the ionized gas, on the strength of the magnetic field and on the radio wavelength. If the average ionic density between us and the source of radio emission is known then the average magnetic field can be calculated. The Faraday rotation method is particularly useful when applied to radio signals from pulsars. First, their radio emission is strongly polarized and therefore easy to measure: second, as we discussed in Chapter 5, we can deduce the mean density of ionized gas between us and the pulsar from the differences in the arrival times of the radio pulses at different wavelengths. The Faraday method gives a mean magnetic field near the Solar System of 1.6 microgauss, – a little less than that indicated by the radio emission. Part of the difference may be that the Faraday method only tells us about the magnetic field in those parts of space which are ionized, whereas synchrotron emission comes primarily from regions of space where the magnetic field is stronger than average. Also, the Faraday effect tells us only about that part of the field that points towards us or away from us, and nothing about the field at right angles to our line-of-sight.

The Zeeman effect complements the Faraday rotation method in that it works for regions of space that are neutral rather than ionized. When placed in a magnetic field, the electron in an atom of neutral hydrogen can be either accelerated or decelerated in its orbit around the nucleus. Its energy is thereby increased or decreased slightly. The details of its behavior can be understood only with the aid of quantum mechanics, but the overall effect is comparatively simple; the 21-cm line becomes split into two lines with a separation in wavelength that is proportional to the strength of the magnetic field. The effect is very small, producing separations that are less than the broadenings produced by the Doppler effect, but the effect has been used successfully to measure the magnetic fields of comparatively dense clouds. A similar effect in the 18-cm lines of the hydroxyl (OH) molecule has been used in clouds where the hydrogen is predominantly molecular. The Zeeman effect is the only technique that allows for a measurement of the field at a specific location; other methods all give us some kind of average field between us and some distant point, but if we can isolate a particular cloud by its Doppler shift we study the Zeeman effect in that cloud by itself. An important result of the Zeeman studies is the discovery that some dense clouds have magnetic fields many times higher than the average values for interstellar space, a result that lends support to the idea that frozen magnetic flux has been amplified as clouds have become compressed. In Chapter 12 we will return to the subject of magnetic fields in dense clouds, and see how they are crucial in moderating the rate at which new stars are formed in the Galaxy.

10 The origin of interstellar matter

The interstellar medium is in a state of perpetual flux. It is continually diminished as stars are formed from it and replenished as they eject material back into it. Its history is therefore closely linked to the history of the Galaxy as a whole. In this Chapter we try to answer three questions. First, where do the various elements that make up our Universe come from? Second, how do they get into the interstellar medium? Third, how do some of them get locked into grains?

10.1 The origin of the elements

There is now fairly good agreement among astronomers that the Universe, and all matter in it, originated in an instantaneous 'Big Bang', 10–20 billion years ago. Since that time the Universe has been steadily expanding, as witnessed by the fact that the distant galaxies are all seen to be moving away from us at large velocities. In the first few seconds after the Big Bang, individual atoms did not exist, and the Universe consisted only of an immensely concentrated inferno of matter and energy in the form of elementary particles such as electrons, protons, neutrons and photons. As the Universe expanded and cooled over the next few minutes **nucleosynthesis** began as protons and neutrons collided with each other with sufficient violence that thermonuclear reactions took place throughout all of space. Initially they combined in pairs to form **deuterium** or 'heavy hydrogen' nuclei consisting of one proton and one neutron. Some of these deuterium nuclei still exist in interstellar space, in a concentration of one deuterium atom to every 100 000 regular hydrogen atoms. Most of them, however, quickly combined in pairs to form helium nuclei, consisting of two protons and two neutrons. Small numbers of lithium nuclei (three protons and four neutrons) and of the light isotope of helium (two protons and one neutron) were also formed during this era, but by about 12 minutes after the Big Bang the Universe had expanded so much that collisions between particles became too infrequent and insufficiently violent for fresh nuclear reactions to take place. The ratio of one helium atom to every ten hydrogen atoms became fixed, except that for nearly a million years after the Big Bang the temperature of the gas in the Universe was so high that all the atoms were ionized rather than neutral. The Big Bang produced only a few atoms of lithium for every 10 000 000 000 of hydrogen.

Hydrogen, helium and lithium are the only elements whose origin can be traced back to the Big Bang. All the other elements were formed much later, after the hot primordial gas had cooled and gravitational forces had concentrated it first into galaxy-sized gas clouds and then into individual stars. It is these stars, and those which came later, that have been responsible for the creation of all the elements heavier than lithium. In the cores of stars the gas pressures and temperatures can rival those at the time of the Big Bang, and nuclear reactions can occur with great intensity, building more complicated (i.e. larger atomic number) atoms out of simpler ones.

Paradoxically, the most common series of thermonuclear reactions in the Galaxy is relatively unimportant for the production of new elements. These are the reactions that occur in main-sequence stars, including the Sun, and result in the conversion of hydrogen into helium. The quantities of material being processed are enormous. In our Sun alone, 600 million tons of hydrogen are converted to helium every second in a process that is duplicated in some 10^{11} stars in our Galaxy. Vast though the amounts of helium produced inside stars are, the total amount synthesized in the past 16 billion years is far smaller than the amount that was produced in the Big Bang. Main-sequence stars are therefore not very important sources of new elements; it is in **evolved stars** – those at the stages beyond the main sequence – that the more interesting nuclear reactions take place.

To understand how elements are synthesized in evolved stars we need to learn a little more about how stars work. The gas in a star continually adjusts itself so as to maintain a number of balances throughout its interior. Chief among these is the balance between gas pressure forcing a star to expand and gravity holding it together. Also of great importance is the balance between the amount of thermonuclear power generated in the core of the star and the amount radiated into space from its surface. In most stars, thermonuclear reactions can only occur in a small region around the center of the star, but any alteration in the production rate of nuclear energy in the center will affect the gas pressure and energy flow throughout its interior. Changes in the nuclear reactions taking place in the interior of a star can therefore lead to substantial changes in its temperature and size. These changes at the star's surface usually do not bear a simple relationship to those at its core, but they can be predicted by careful calculations. These computer calculations are often referred to as **stellar models** because they effectively reduce the size of a star to that of a desk-top computer, and cut its evolution time from billions of years to a few minutes.

The mass of the Sun (2×10^{27} tons) provides it with enough hydrogen fuel to last about 10^{10} years – about double its present age. The model calculations predict that after this time the central regions of the star become choked with helium. Like a fire spreading in a forest, hydrogen burning now forms a shell that moves outward through the star leaving behind a core of almost pure helium. Gravitational forces cause this core to contract and get hotter. Eventually the temperature and density reach the point that a new series of nuclear reactions start; helium nuclei combine to form more complicated elements, releasing thermonuclear energy as they do so. The most important products are the main isotopes of carbon and oxygen, which are formed out of the **fusion** of three and four helium nuclei respectively.

During these later stages in a star's life nuclear reactions take place very rapidly in the star, so it has to radiate away much more power than it did when it was burning hydrogen in its core. The extra heat in the star causes it to expand to many times its earlier size. At the same time its surface temperature (but *not* the temperature of the core) decreases. The name **red giant** is given to these large, comparatively cool stars. When the time comes for the Sun to become a red giant it will grow so large that it will engulf the planet Mercury.

The helium-fueled stage of a star lasts only for about 1% of the hydrogen-fueled stage. After this time conditions in the core of the star change again and several new generations of elements are formed in rapid succession. Particularly important are reactions in which a helium nucleus is fused into an existing nucleus. For this reason, many of the more abundant elements in the Universe have atomic weights which are exactly divisible by four. Examples include neon, magnesium, silicon, sulfur, argon, calcium, iron and nickel, as well as carbon and oxygen and helium itself. Other elements are formed from these nuclei mainly by adding or subtracting individual protons, neutrons and electrons. Nitrogen is a special case; it is made from carbon and hydrogen in the cores of main-sequence stars as a by-product of hydrogen fusion.

All the nuclear reactions we have discussed up till now have been **exothermic**; in other words they cause nuclear energy to be given off, primarily in the form of gamma rays. Thermonuclear fusion in red giants and other kinds of stars can account for all the elements with atomic numbers up to about 28. Elements heavier than iron or nickel can be synthesized only by **endothermic** reactions in which energy is absorbed into the nucleus. In a stable star no suitable source of energy exists. It is therefore possible for a star to reach a stage where it runs out of nuclear fuel, and is no longer able to maintain the gas pressure that supports it against its own gravity. The ultimate fate of such a star depends on how much mass it has when its fuel runs out. Stars like the Sun shrink to form a **white dwarf** star which eventually cools and fades into darkness. Stars with large masses, however, end their lives with a spectacular explosion called a **supernova**. In a supernova explosion energy is released as a result of the sudden collapse of the star's core. Part of the liberated energy goes into expelling material into space, and part into the creation of new elements. These explosions occur only a few times a century in our Galaxy, but are together responsible for the synthesis of all the elements heavier than iron, including many that we come across in daily life, like copper, mercury, gold, iodine and lead. Most of the elements which are produced in supernovae have small cosmic abundances, and very few have been directly detected in the interstellar medium.

10.2 Mass loss from stars

Since the interstellar medium contains plenty of heavy elements – i.e. those with atomic numbers ≥ 6 – we know that it cannot simply be the remnant of gas left over from the Big Bang. At least some of the gas must once have been part of the interior of a star, and been subsequently expelled from it. Here we look at how stars replenish the interstellar medium and enrich it with heavy elements.

The amount of gas that a star gives off depends on many factors. Chief of these are its mass and its age. During the period soon after their birth, most stars go through a stage of adjustment that involves ejecting material from their surfaces in the form of a powerful **stellar wind**. As we will discuss in Chapter 11, winds from young stars have a significant effect on the evolution of both the star itself and on the cloud that surrounds it. On a galactic scale, however, mass loss from young stars is not very significant since it involves material that was only recently extracted from the interstellar medium and which has not had its chemical composition altered by its brief exposure to life inside a star.

During the main-sequence stage of its life, when a star is burning hydrogen to make helium, its stellar wind continues to blow, though with less intensity than before. In the case of the Sun, the wind amounts to some 2 million tons per second and flows past the Earth in a fast moving stream of ions called the **solar wind**. Although large in human terms, the rate of loss of matter from the Sun amounts to less than 1% of its total mass over its whole

10^{10} year lifetime. Larger stars have stronger winds, and some bright OB stars eject more hydrogen into space than they convert into helium.

For the majority of stars, the most intense period of mass loss comes in the late stages of their evolution, starting from the time they become red giants. The reason is partly gravity; if a star expands to 10 times its original radius, the force of gravity on its surface is reduced by a factor of 100, and much less energy is required to eject atoms into space. The mass loss rates from evolved stars can be enormous – up to 10^{22} tons a year. The rapid loss of material can have drastic consequences for the outward appearance of a star as well as on the course of its evolution. Normally there is not much mixing in a star. The elements produced by nuclear fusion remain in the star's interior, while its outer layers continue to have much the same chemical composition as the material out of which it was formed. If enough mass is lost from a star, however, its inner layers – rich in helium, carbon and other heavy elements – may become exposed and ejected into space. This is one way that the interstellar medium becomes enriched in heavy elements. As we will discuss later in this chapter, the mass loss process may culminate in the production of a planetary nebula, or a supernova, depending on the mass of the star.

How are stellar winds from evolved stars observed? One important method is to observe the emission and absorption lines produced in the stellar spectrum by the material flowing out from the star towards us. Infrared, visible and ultraviolet data are all useful for this purpose. Absorption lines produced in outflowing gas are distinguishable from the lines produced in the star's atmosphere by their different Doppler shifts. From the wavelengths and strengths of the spectral lines an astronomer can, in suitable cases, estimate the speed, density, and composition of the gas in the stellar wind.

Another method involves observing radio or millimeter-wavelength emission lines from molecules in the gas moving out from the star. The most useful molecule is carbon monoxide (CO). In some cases the gas extends far enough from the star that it can be mapped using a radio telescope. The main disadvantage of the method is that CO comprises only about 1 part in 10^4 of the gas in the outflow, and is formed by chemical reactions that are by no means fully understood. The other molecule that is widely observed in the outflows of evolved stars is the hydroxyl molecule (OH), which is unstable in the Earth's atmosphere but fairly widespread in the low-density reaches of interstellar space. Around some stars the emission from the 18-cm wavelength spectral lines of the OH molecule is so intense that some kind of maser process (see Chapter 9) must be taking place. The stars around which maser emission is observed are all variable stars as well as being giants or supergiants; their diameters, and hence their light and their radio emission oscillate over periods of months.

The third way in which stellar winds are detected in evolved stars is by their infrared emission. As we will discuss later in this chapter dust grains can condense in the outflowing gas of a cool star. When this happens, the grains absorb light from the star in the same way as interstellar grains absorb starlight in the interstellar medium. The amount by which the star is dimmed is usually small, but the excess radiation emitted by the heated dust at infrared wavelengths is easily detected. The stars μ Cephei and IRC +10216 (Figure 7.7) are examples of stars with circumstellar shells formed in their stellar winds.

The ways that stellar winds are generated are not at all well understood. The wind has to be started by some kind of disturbance within the star that causes matter to be thrown outward from its surface. Two kinds of disturbance that may work are slow pulsations of the whole star and changing magnetic fields near the surface; the latter process can be seen in action, albeit at a fairly low level, on the surface of the Sun. Once the wind is on its way it may get an extra push from the **radiation pressure** of the light from the star itself. Radiation pressure is an electromagnetic force that is exerted on matter by photons. Under most circumstances it is too weak to be important but it is useful for accelerating material that has already moved far enough from the star's surface for dust grains to condense.

10.3 Planetary nebulae

The ultimate fate of a star after its various post-main-sequence phases depends mainly on its mass. Small stars just fade away, but larger ones complete their lives with a flourish forming either a planetary nebula or, in the case of the most massive stars, a supernova.

A planetary nebula arises when the wind from a red giant becomes so intense that all but the dense core of the star is ejected into space. The star that emerges from this eruption no longer bears any resemblance to a red giant. It is now small and intensely hot. Because it is so hot it emits copious quantities of ultraviolet radiation, which ionize the gas previously ejected, rendering it visible as a nebula of glowing gas that surrounds the star. The name planetary nebula is misleading, and rose historically from the visual resemblance between certain small circular nebulae and Solar System planets such as Uranus when viewed in the telescopes of the late eighteenth century.

Planetary nebulae are somewhat like H^+ regions, in that both consist of gas ionized by a hot star. They have similar visible-wavelength spectra consisting mainly of recombination lines of hydrogen and helium together with forbidden lines of ionized oxygen, nitrogen, neon and sulfur. The central star of a planetary nebula is usually hotter than that of an H^+ region, with a surface temperature in the range 50 000–100 000 K rather than 25 000–50 000 K; the higher stellar temperatures mean that many more highly excited ions, such as Ne^{4+} and in particular, He^{++} are found in the ionized nebula. The radio and infrared properties of planetary nebulae also resemble those of H^+ regions. Collisions between electrons and ions produce radio-wavelength free–free continuum emission while dust grains absorb ultraviolet photons from the star and from the nebula, and re-emit the energy as infrared radiation.

It is usually quite easy to distinguish a planetary nebula from an H^+ region. First, it is generally highly symmetric in appearance, with a single star at its center, unlike an H^+ region which is usually ragged, and may be excited by a group of hot newly-formed stars. Second, it is likely to be isolated from other interstellar clouds, unlike H^+ regions which tend to congregate around each other and around dense molecular clouds. Third, it has a much smaller total mass of gas – less than 1 solar mass – than most H^+ regions.

There are more than 1000 planetary nebulae known. The best studied ones have angular sizes between 10 arcseconds and several arcminutes, but planetaries can range in size from giant objects such as the Helix Nebula (Figure 5.4), which is about 1 degree in diameter, to objects so small that, were it not for their emission-line spectra, they would be mistaken for stars. Planetary nebulae exhibit a variety of shapes, but most often have the appearance of either a ring or an hourglass. The different shapes probably arise from variations in the way that the gas was initially expelled from the star. The speed of rotation of the star, its radiation pressure and its magnetic field may also be important.

There are two pieces of evidence to show that planetary nebulae are expanding away from their central stars. First, photographs and radio

Figure 10.1. The Dumbbell Nebula. The glowing ionized gas has been ejected from the star at the center of the nebula.

images of certain planetary nebulae obtained several years apart show directly that they are slowly increasing their diameters. Second, the Doppler shifts of the light emitted from different parts of the shell show that in many nebulae the gas is expanding away from the star at around 20 km s^{-1}, a speed that is quite similar to that of the wind from a red giant star. At this speed it would take a planetary 25 000 years to expand to a diameter of one parsec.

The transition from a red giant to a planetary nebula is clearly a dramatic event in a star's life. Is there anywhere in the Galaxy where we can see the process taking place? Recently a number of objects have been discovered which appear to be planetary nebulae in formation; they are termed **bipolar nebulae** as a result of their visual appearance. The structure of a typical bipolar nebula is shown in Figure 10.2. The star itself is obscured by a ring of thick cool gas and dust, the heated inner part of which emits strong infrared radiation. The ring is comprised of matter expelled by the star; its shape probably results from the star's rotation. Along the axis of the ring, where the density is lower, some starlight escapes to illuminate the diffuse patches of dust that we see as reflection nebulae. If the star is hot enough it may ionize a small zone close to its surface, but most of the gas in the bipolar nebula is molecular. We anticipate that as the ring of matter expands outward it will become more transparent and, as the hotter core regions of the erupting central star become exposed, a wave of ionization will spread out through the material surrounding the star creating the planetary nebula. The timescale for these changes is very hard to estimate, but is probably of the order of a few hundred or few thousand years.

Once the planetary nebula is fully formed its central star will cease to eject new matter. The gas in the nebula therefore has no means of replenishing itself and drifts off into space, gradually losing its identity as it blends with the diffuse interstellar medium. The central star, meanwhile, with no

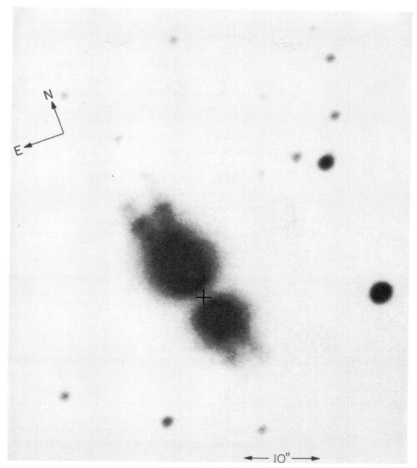

Figure 10.2. The Egg Nebula is a small bipolar nebula that represents an intermediate stage between a red giant and a planetary nebula. The star itself is hidden behind the dust lane in the center of the nebula.

nuclear fuel left to burn, slowly cools over a period of some 10^5 years becoming first a **white dwarf**, and, finally, an inert **black dwarf** star, fading into invisibility as it does so.

The importance of planetary nebulae to our story is threefold. First, they play a major role in the recycling of matter from stars back into the interstellar medium. Second, the white dwarf stars they leave behind produce so much ultraviolet radiation that they contribute significantly to the ionization of the diffuse interstellar medium. Third, the ease with which they can be observed, particularly by spectrographs, resulted in their providing much of the data by means of which the behavior of atoms and ions in interstellar space were first understood in the early part of the twentieth century.

10.4 Supernovae

White dwarf stars are remarkably dense objects. Roughly speaking, they contain the mass of the Sun in a volume the size of the Earth. The matter in a white dwarf is quite unlike matter found on Earth or in the interstellar medium. Electrons are no longer tied to particular atomic nuclei, but take the form of a dense 'sea' of particles that are pressed close together by the gigantic gravitational forces within the star. For reasons that can be understood only in the context of the laws of quantum mechanics these confined electrons exert a **degeneracy pressure** that prevents the star from collapsing under its own weight. This degeneracy pressure differs from normal gas pressure in that it does not depend on a supply of heat; consequently, as a white dwarf star cools towards a black dwarf, its size does not change.

White dwarfs cannot exist with a mass greater than about 1.4 $M_\odot$, because the degeneracy pressure in such a massive object would be insufficient to withstand its gravitational forces. When the nuclear fuel runs out in a star with a mass greater than this, therefore, a quite different series of events takes place. Its core becomes unstable and suddenly collapses in less than a second to become a **neutron star**, an event that triggers a **type II supernova explosion**. Neutron stars are even more remarkable than white dwarfs; they have a diameter of only a few kilometers and a density similar to that found within the nucleus of an atom. One teaspoonful of neutron star material would weigh some 10 billion tons. The formation of a neutron star involves the release of vast amounts of gravitational energy as the core of the star falls inward towards its center. Simple physics tells us that since the diameter of a neutron star is 100 times smaller than that of a white dwarf, 100 times as much energy is released by its formation. Part of this energy is absorbed by nuclear reactions that create a vast array of chemical elements that can only be made during a supernova explosion. Part goes into blowing the outer layers of the star, including some of the newly-synthesized elements, into space at speeds of thousands of km s^{-1}. Part goes into turning the escaping gases into a brilliant fireball whose power output, for a few exciting days, can match that of a whole galaxy.

It is estimated that supernova explosions happen in our Galaxy about once every 20 years. Most are hidden from us by obscuring dust, however, and the last one seen was in 1672. All the supernova explosions that have been studied since then, including the famous 1987 event in the Large Magellanic Cloud, have occurred in other galaxies. Although the stellar

Figure 10.3. The Crab Nebula. This nebula in the constellation of Taurus is the remnant of a supernova that exploded in the year 1054. In the center of the nebula is a pulsar which flashes 30 times per second.

collapse and its ensuing nuclear reactions are over in a fraction of a second, the visible manifestations of a supernova last longer. The hot expanding gas cloud reaches its maximum brilliance in a few days, then fades over a period of a few months as it cools and dissipates into space. The interaction of the expelled gases with the interstellar medium is such that they maintain a recognizable identity for several thousand years; during this time they are referred to as a **supernova remnant** – often abbreviated to SNR. Over a hundred SNRs are known in our Galaxy, most dating from supernova explosions that occurred in prehistorical times.

The most famous supernova remnant in our Galaxy is the **Crab Nebula**. It is the remnant of a supernova explosion which was observed by Chinese astronomers in July 1054. This supernova was so bright that it was visible in full daylight for three weeks. The nebula is now too faint to be seen with the naked eye, but is visible as a fuzzy patch with a small telescope. It is about 2000 pc away, about 4 pc in diameter, and growing steadily. The star which exploded to form the nebula, and which is now a neutron star, can be identified by the fact that it is a **pulsar**, flashing on and off like a lighthouse. The light we see is not from its neutron star surface but from charged particles trapped in the star's magnetic field and forced to rotate with it 30 times a second. The pulsar is crucial to the nebula since it is the reservoir for the power the nebula radiates into space. A neutron star spinning 30 times a second acts like a giant flywheel and contains a vast reserve of kinetic energy. From the measured rate at which the pulsar is slowing down we can calculate the rate at which the energy is being lost by the pulsar. We find that this power is equal to the total power radiated by the nebula (mainly at X-ray wavelengths). By some process that we do not yet understand the rotating neutron star is able to accelerate streams of electrons to speeds close to that of light. The interaction of these relativistic electrons with the surrounding magnetic field gives rise to synchrotron radiation, which is the source of the diffuse whitish glow seen from the central parts of the nebula. Synchrotron radiation is generated over a very wide range of energies, and causes the Crab Nebula to be one of the most prominent objects in the sky at both radio and X-ray wavelengths. The filaments which give the nebula its characteristic crab-like shape, and which contain most of its mass, are ionized gas. The ionization is maintained not by a star, but by ultraviolet photons generated by the synchrotron process.

Although it is the most-studied and best-understood SNR in the Galaxy, the Crab Nebula, with its spinning pulsar, is not typical. In most supernova remnants the radiated energy derives from the momentum of the expanding gases, rather than from the rotation of a central neutron star. With no spinning neutron star to re-accelerate them, the relativistic electrons quickly lose much of their momentum, and their synchrotron emission is detectable only at radio wavelengths.

Most SNRs are hollow, with a shape something like the thin skin of a peeled orange. The strongest emission comes from the edges where their expanding gases collide with the surrounding interstellar medium. Vast quantities of kinetic energy are released as the fast moving gases are squeezed and forced to slow down. Much of this energy is converted to electromagnetic radiation, by several quite distinct processes. First, because the compression of the gas raises its temperature to millions of degrees, supernova remnants are powerful sources of thermal X-rays (Figure 10.4). Second, because the compression process boosts any magnetic field that is locked into the colliding gases, supernova remnants are particularly strong sources of radio-wavelength synchrotron emission (Figure 10.5). Thirdly, because of the greatly enhanced rate of collisional excitations, supernova remnants may radiate emission lines at visible and ultraviolet wavelengths,

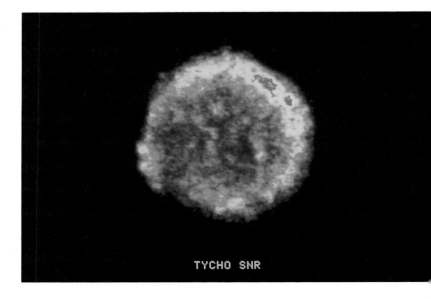

Figure 10.4. The Tycho supernova remnant as seen at X-ray wavelengths by the Einstein satellite. The emission comes from hot gas heated to temperatures of millions of degrees.

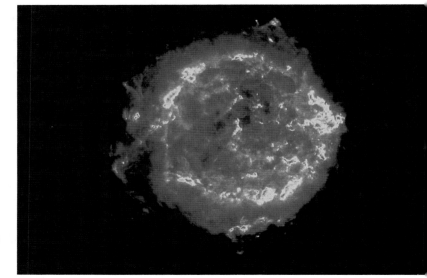

Figure 10.5. The Cassiopeia supernova remnant as seen at radio wavelengths. The radio emission comes from relativistic electrons that are generating synchrotron radiation as they spiral around the magnetic field in the supernova remnant.

causing them to appear as delicate nebulae (Figure 10.6). Some supernova remnants show faint visible-wavelength emission lines of highly ionized iron atoms such as Fe^{9+} or Fe^{13+}. These ions are characteristic of coronal gas (see Chapter 11).

At radio and X-ray wavelengths supernova remnants are among the most prominent objects in the sky. At visible wavelengths, on the other hand, they are usually faint and hard to observe, but because they can be studied spectroscopically, they yield a great deal of information. Some of the faint filaments associated with the Cassiopeia supernova remnant are particularly interesting. Velocities of up to $10\,000$ km s^{-1} have been measured by the Doppler shift, indicating that some material is still moving at close to the speed it left the exploding star 300 years or so ago. Other filaments have been found in which the abundance of oxygen as compared to hydrogen is thousands of times greater than normal. The gas in these

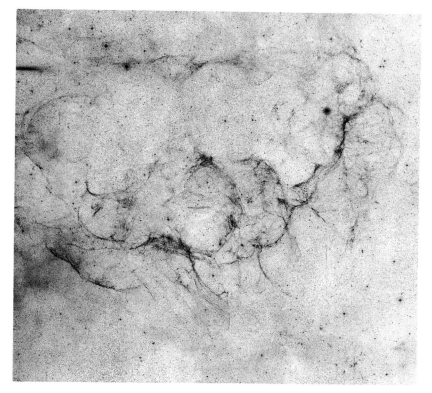

Figure 10.6. The Vela supernova remnant in a negative photograph. The wisps are comprised of interstellar matter that has been swept up and compressed by the force of the supernova explosion.

filaments is presumably derived from deep in the core of the exploded star, where its hydrogen and helium had all been consumed by thermonuclear reactions.

The total amount of material returned to the interstellar medium by supernova explosions is small compared with that from red giant stars and from planetary nebulae. Supernovae are crucial in two ways, however. First, they are the main means by which the interstellar medium becomes enriched with heavy elements. Second, as we shall see in Chapter 11, because of the chaos they produce in their surroundings, supernova remnants are largely responsible for the interstellar medium having the complicated structure that it does.

10.5 Tools of the trade: X-ray telescopes

Like ultraviolet waves and gamma rays, X-rays can be observed only from above the Earth's atmosphere. Experiments with small rocket-borne telescopes in the 1960s convinced astronomers that it was worth launching a number of major satellite experiments in the 1970s. Some were designed to monitor X-rays from the Sun, some to make X-ray surveys of the sky, and some to study individual objects in detail. The satellite which produced the most spectacular results was the Einstein Observatory which was launched in 1978 and flew successfully for $2\frac{1}{2}$ years. This satellite was capable of seeing finer details in X-ray sources than any other telescope before or since (Figure 10.4). Unfortunately X-ray satellites die rather quickly, usually because they run out of fuel to control the pointing of the satellite. As a result, X-ray astronomy is a stop-and-start exercise; much activity follows the successful launch of a new satellite, but long fallow periods exist when the absence of a working satellite has meant that no X-ray astronomy is possible at all.

X-ray telescopes share some of the features of gamma ray telescopes. They contain electronic instruments that count the individual X-ray photons as they reach the spacecraft. Information about each photon is then radioed back to Earth. Although conventional lenses and mirrors do not work at X-ray wavelengths, special techniques have been developed that can produce focused images that are nearly as good as those of optical telescopes. The mirrors for X-ray telescopes are very hard to make and, as a result, the effective diameters of X-ray telescopes are usually measured in centimeters rather than meters (as in optical astronomy) or tens of meters (as in radio astronomy).

Ordinary stars, planets and galaxies do not give off much in the way of X-rays. The objects studied by X-ray astronomers therefore tend to be somewhat exotic. Besides supernova remnants, the important sources of X-ray emission include black holes, quasars, closely-separated binary stars and clusters of galaxies. The two main physical processes that give rise to X-rays are basically the same as those that produce radio waves. These are free–free emission from ionized gases, which was described in Chapter 5, and synchrotron emission from high-speed electrons which was described in Chapter 9. Both mechanisms produce mainly continuum emission, which provides one reason why so much of X-ray astronomy is based on broad-band as opposed to spectroscopic observations. A few emission lines have been observed in the X-ray spectra of hot gases but at the present time X-ray spectroscopy is much less advanced than spectroscopy at ultraviolet or visible wavelengths.

Most of the pioneering work on X-ray astronomy was carried out by American astronomers, but constraints on NASA's budget have led to a hiatus in US X-ray astronomy since the demise of the Einstein satellite in 1981. Countries that have successfully flown X-ray telescopes now include the UK, the Netherlands, Japan, and the USSR as well as the European Space Agency consortium. The launch of the X-ray satellite ROSAT by West German scientists in 1990 puts Europe at the leading edge of research in this area, but US astronomers hope to regain the initiative when the Advanced X-ray Astronomy Facility (AXAF) is ready for launch sometime in the mid 1990s.

10.6 Novae and other binary systems

More than half of all stars are members of **binary systems** in which two stars are bound together by their mutual gravitational attraction, and revolve in stable orbits around each other. Usually the stars are so far apart that each evolves in its own way through its red giant stage to its death as either a white dwarf or a neutron star. In a fascinating minority of binary systems, however, the stars are separated, for at least some of their lives, by little more than their diameters. Under these circumstances some very interesting things can happen if one star starts to eject matter into space.

The word **nova** is used to describe the phenomenon whereby a faint star explosively brightens by a factor of a thousand or more and then fades to its original luminosity over the subsequent few months. Novae are intrinsically less powerful than supernovae, but occur much more frequently. Several new ones are discovered each year – often by skilled amateur astronomers. It is now believed that novae consist of a close pair of stars, one of which is a main-sequence star and the other a white dwarf. Gas that emerges as a stellar wind from the main-sequence star is attracted by gravity towards the white dwarf star and collects on the surface. This gas is rich in hydrogen – unlike the white dwarf, whose hydrogen has all been converted to heavier elements. Because it has such a small diameter, the force of gravity at the surface of the white dwarf is gigantic, resulting in a rapid build-up in the gas pressure there. Eventually the pressure in the newly-acquired layer of hydrogen becomes so large that a thermonuclear explosion takes place in which hydrogen is converted into helium, and the envelope of gas is ejected violently into space. In some novae this gas becomes apparent as a small nebula some years after the event. The whole process can then be repeated.

The combination of a main-sequence star and a white dwarf can also give rise to a **type I supernova**. The difference is that in this case the extra weight of the accreted matter builds up the pressure inside the white dwarf. Eventually this pressure becomes enough to set off explosive thermonuclear fusion reactions in the helium or carbon nuclei that make up the bulk of the white dwarf star. Type I supernovae can be as bright as type II supernovae, but they do not lead to the production of any elements heavier than iron.

10.7 Formation and destruction of dust

So far in this chapter we have established where the elements came from and how they got into interstellar space. We still need to understand how some of the atoms come to be bound up into grains. In particular we would like to confirm the idea presented in Chapter 7 that the reason why some heavy elements have a low gas phase abundance is that they have been incorporated into dust grains.

A key to the question of the origin of grains is the fact that red giant stars that are expelling matter into space are strong sources of infrared emission. From the strength of the infrared radiation we can estimate how thick the dust must be in the vicinity of the star. We find that in most cases there is far more dust than would be expected in a random piece of interstellar space. Red giants are too old to retain any of the dust from the cloud they were born with and do not have the gravitational pull to attract a retinue of grains out of the general interstellar medium. The source for the grains must therefore be the star itself. Since the star's interior and atmosphere are too hot for grains to survive, they must form out of the stellar wind of the star. The process is essentially one of condensation. As gas moves away from the heating influence of the star it cools from a temperature of several thousand degrees to some tens of degrees. As specific temperatures are reached by the cooling gas, different solid materials condense out of it, just as soot forms in a candle flame, and snow crystals condense in moist air when its temperature drops below the freezing point of water.

Predicting which solids condense out of a flow of cooling interstellar gas is a complicated chemical problem, partly because the density of the gas is changing even more rapidly than its temperature as the gas expands away from the star. Fortunately there is an interesting piece of chemistry that drastically simplifies the problem. As we have noted before, the carbon monoxide (CO) molecule is extremely stable. The affection of oxygen and carbon atoms for each other is so strong that the formation of CO dominates the chemistry of the outflowing gas. If, as in most stars, oxygen atoms outnumber carbons, then essentially all the carbon atoms in the outflowing gas are incorporated into CO molecules and none are available for making dust. In stars where carbon atoms are the more numerous the dust grains can contain no oxygen atoms. We therefore have to consider **oxygen-rich** and **carbon-rich** stars separately.

As gas flows out from a star the first solids to condense are those which vaporize at the highest temperatures. The process starts with the metal tungsten at 2000 K, but this element is so rare that the amount of tungsten dust formed is negligible. In normal (oxygen-rich) stars the next grains to form are oxides and silicates of titanium, aluminum, and calcium, at temperatures of around 1500 K. Because these condense close to the star where the gas density is high the grain formation process is very efficient. Consequently these three elements are among the most strongly depleted from the interstellar gas (Chapter 6). Most of the magnesium, and the rest of the silicon atoms condense into silicate crystals near 1200 K, where the density is somewhat lower and the grain formation efficiency is less. Consequently they are not as strongly depleted as metals in the first group. Although oxygen is a necessary constituent of silicate crystals the gas in a normal stellar wind contains so many more oxygen atoms than silicate or magnesium atoms that the formation of silicate grains does not cause significant oxygen depletion from the gas phase.

About 10% of stars have atmospheres which contain more carbon atoms than oxygen atoms, as a result of a stirring process that lifts processed material from the core of the star towards its surface. No oxides or silicates form in the wind from a **carbon-rich** star. Probably the most important process is the condensation of carbon atoms to form crystals of graphite at around 1800 K. Infrared spectra of carbon-rich giant stars also show an emission feature attributable to silicon carbide, which is never seen from oxygen-rich stars, confirming that the chemistry of the outflowing gas is very different in these two kinds of object.

There is some evidence that dust grains form in the material ejected from novae and supernovae. A few months after the 1980 explosion of a supernova in the galaxy NGC 6946, astronomers detected infrared emission from dust at a temperature of around 800 K. A similar phenomenon was seen following the appearance of a bright nova in the constellation Cygnus in 1978. The emission suggests strongly that dust grains condense in the ejected material, but it is possible that the dust was there before the explosive event, and was reheated by it.

Not all grains are formed in the wind of a red giant star. We know from infrared spectroscopy that there are ice crystals present in molecular clouds, but not in the general interstellar medium. What probably happens is that in the cool darkness of the molecular cloud water molecules, together with those of other simple chemicals like methane (CH_4) and ammonia (NH_3), condense to form icy coverings around the mineral grains. These layers provide excellent opportunities for the formation of more complicated compounds as the material in the grains is chemically reprocessed.

The problem of the formation of interstellar grains is linked to that of their destruction. Grains can be damaged or destroyed by several processes, including evaporation, and collisions with other fast-moving grains or atoms. These destruction processes are particularly powerful where gas clouds are colliding at high relative speed. Indeed, it is found that gas associated with interstellar shock waves is less depleted in heavy elements than the normal interstellar medium; the shocked gas has had its heavy elements replenished by the atoms from the destroyed grains. Theoretical studies suggest that it is rather easy for grains to be destroyed in the interstellar medium, so much so that it is becoming difficult to understand how some elements, like calcium, can remain so strongly depleted. The dilemma suggests that perhaps grains are more resilient than we think, or that there may be some additional grain formation process that we have not paid enough attention to up till now.

11 Solving the interstellar jigsaw puzzle

The organization of this book reflects the way in which astronomers usually think about a complicated problem. They like to break it up into smaller pieces which can be tackled individually. This is why this book has separate chapters on topics such as atoms, ions, molecules, dust and cosmic rays. Since we have now introduced all the major constituents of the interstellar medium we can start to ask how they all fit together in the Galaxy. The problem is like a jigsaw puzzle in more ways than one. We want to know how the various kinds of clouds and intercloud gases fit together to fill the space between the stars, but we also need to understand how each form of interstellar matter is affected by its neighbors. Is gas heated by dust, or dust by gas? How it is that gases with temperatures as low as 10 K and as high as 1 000 000 K can coexist in the same galaxy? Why has not the gas in our Galaxy smoothed itself out into a bland calm atmosphere? Before attempting to answer these questions let us pause to review the various kinds of interstellar gases that we have met in this book up till now.

11.1 Towards a taxonomy

Astronomers are much less prone to classifying their discoveries than biologists. Consequently there is no rigorous and generally accepted taxonomy of interstellar phenomena. One astronomer's 'giant molecular cloud' is another's 'molecular cloud complex'. One person's 'H^+ region' is another's 'gaseous nebula'. Nevertheless, some useful classification can be made of the phenomena we have met so far in this book. In the scheme used below, the primary classification characterizes the recent history of the gas; the secondary classification characterizes its temperature. A further sub-classification, based on the physical state of the hydrogen, is necessary in two cases.

Gas recently ejected from stars

Circumstellar shells are generated by winds from the surfaces of red giant stars.

Planetary nebulae are formed when circumstellar gas is ionized by the hot remnant of the star it surrounds.

Supernova remnants result from the explosion of an unstable star.

Gas associated with the births of stars

Dense molecular clouds have temperatures around 10 K, rising to 50 K near their edges or near embedded new stars.

H$^+$ regions are ionized nebulae around newly-formed OB stars. Their gas temperatures are around 10 000 K.

Gas in the diffuse interstellar medium

Cool clouds. These have temperatures of around 80 K and particle densities of the order of 10–100 atoms cm^{-3}. We must distinguish between:

Molecular gas which is seen by its ultraviolet absorption lines, and

Atomic gas which is seen by its 21-cm emission line as well as by its ultraviolet absorption lines. A single cool cloud may have portions that are mainly atomic and portions that are mainly molecular.

Warm gas has a temperature near 8000 K, and a density of around 0.1 atoms cm^{-3}. Warm gas itself is a mixture of:

Neutral gas which is seen by its 21-cm emission, and

Ionized gas which emits diffuse light and causes delays to pulsar radio signals.

Coronal gas has a temperature of around 10^6 K and a density of less than 10^{-2} atoms cm^{-3}. It emits X-rays and can also be detected by ultraviolet absorption lines of highly ionized atoms.

We discussed gas recently ejected from stars in Chapter 10. We will be dealing with phenomena connected with the birth of stars in Chapter 12. It is with the third main category, the gas in the diffuse interstellar medium, that the rest of this chapter is concerned.

11.2 Heating and cooling of the interstellar medium

It is the natural tendency of any gas to spread out and fill the volume available to it. The atoms in a cool cloud have random velocities of about 1 km s^{-1}; this is fast enough to take them virtually anywhere in the Galaxy during the 16 billion years since its formation. Why, then, have all the density fluctuations in the interstellar medium not smoothed themselves out?

We can obtain a clue to this mystery if we compare the temperatures and densities of the cool, the warm and the hot gas. We find that high temperatures are associated with low densities, and vice versa. More specifically, we find that the gas pressure, which is proportional to the product of the density and temperature (see Appendix G), is about the same in the three phases. This result suggests that the different kinds of gas may be in some kind of pressure equilibrium with each other. If so, how is the equilibrium maintained? This question does not have a simple answer, but many astronomers believe that part of the explanation lies in the way that the interstellar medium is heated and cooled.

An interstellar cloud loses energy in a similar way to an H$^+$ region (Chapter 5); collisions among gas particles cause certain atoms to be excited into upper energy states from which they drop back to the ground state by spontaneous transitions. The downward transitions produce photons which escape from the cloud carrying some of its energy with them. The main difference between H$^+$ regions and other interstellar clouds is in the types of transitions that are excited. In an H$^+$ region at 10 000 K most of the energy is carried away by visible-wave or ultraviolet lines of ions such as O^{++}. In a gas at around 80 K the collisions are much more gentle and the only energy levels that can be excited are those that lie within a fraction of an electron volt of the ground state. The most important of these is an infrared line of ionized carbon, at 156 μm, a wavelength that is, unfortunately, difficult to observe.

The fact that the energy loss from an interstellar cloud depends on collisions has some important implications. Let us imagine that there is some steady heat input, the nature of which we will discuss later, into a region containing diffuse interstellar matter and that, on average, the rate of energy input is balanced by the rate at which energy is radiated out of it. Because collisions occur more frequently when atoms are close together, the energy loss from a dense region of gas is greater than that from a diffuse one. Consequently, if the density of a piece of interstellar gas happens to increase above normal, its cooling rate will rise, causing its temperature to fall. As it cools its gas pressure decreases, causing it to be further squeezed by the surrounding warmer gas. The resultant increase in density leads to a further drop in temperature and sets off a chain reaction that ends only when the gas is so cold that collisional excitation of the 156-μm line no longer occurs efficiently. The opposite effect occurs in patches of the interstellar medium where the density starts off lower than normal. Here, the decreased cooling fails to match the rate at which the diffuse gas absorbs heat. A similar chain reaction starts, causing the gas to become warmer and warmer until some additional form of cooling (such as excitation of the O^{++} lines) stabilizes the gas. The overall result of this process is a spontaneous separation of the gas into two phases, a cool dense one and a warm diffuse one. (At first glance this behavior appears to contradict the common experience that when a gas is compressed its temperature rises. The difference depends on whether or not heat can escape the gas. Warming occurs in cases where heat is trapped in the gas; cooling occurs when heat can be radiated away quickly.)

For the theory outlined in the last paragraph to work there must be a strong source of energy available to heat the interstellar gas. Identifying this heating source is a problem that has not yet been entirely solved. Ordinary starlight will not do the job, since atomic gas absorbs energy only at a few discrete (mainly ultraviolet) wavelengths. The gas cannot be heated by contact with the dust grains, because most interstellar dust is colder than the gas, and, in any case, collisions with dust grains occur too infrequently to facilitate significant exchanges of energy. Two possibilities that have been suggested are X-rays and cosmic rays. Both are capable of transferring energy to the gas, but neither seem to exist in sufficient numbers to heat the gas throughout the Galaxy. The most satisfactory energy source now appears to be **photoelectric heating** by dust grains. The photoelectric effect can be thought of as the photoionization of a dust grain. When an ultraviolet photon is absorbed by a grain it may deposit so much energy into one piece of the grain that an electron is ejected at high speed from the surface. These fast-moving electrons carry with them a significant fraction of the energy of the original photons. As they collide with the surrounding atoms their energy is assimilated into the thermal energy of the gas, thereby raising its temperature. It is almost impossible to determine just how efficient this process is. The main problem is our ignorance of how interstellar grains behave under ultraviolet illumination. In particular, we have essentially no idea how well the photoelectric effect operates for the microscopic PAH grains, which in terms of numbers, though not total mass, probably outnumber all other types of grain. The main argument in favor of the photoelectric effect as a major heat source for the interstellar medium is that it makes use of a known reliable source of energy, namely ultraviolet photons produced by hot stars.

The theory described in this section seems to provide at least a partial explanation for the existence of the two main kinds of atomic gas, the cool clouds and the warm neutral gas. In the early 1970s however, a serious inadequacy of the theory was exposed with the discovery of **coronal** gas in interstellar space. We briefly mentioned coronal gas in Chapter 5. It is now time to look at it in more detail.

11.3 Coronal gas revisited

Coronal gas is a very hot, very thin type of interstellar gas whose existence was revealed by spacecraft observations. The word 'corona' is derived from quite a different branch of astronomy. In its original usage it refers to the thin outer atmosphere of the Sun, which is normally invisible, but which may be seen as a bright 'crown' around the Moon during a total solar eclipse. The temperature of the gas in the solar corona is around 10^6 K, far hotter than the 5800 K of the Sun's visible surface. The word 'coronal' has subsequently become applied to refer to gas of about this temperature in other branches of astronomy. As we shall see in Section 11.6, however, the analogy with the Sun is apt in another way; the Galaxy appears to be surrounded with an envelope of coronal gas, just as the Sun is surrounded by its corona.

The first evidence for interstellar coronal gas came in the form of some unexpected absorption lines in the ultraviolet spectra of certain stars. The crucial feature of these particular lines is the fact that they are produced by interstellar ions which have been stripped of many electrons. One such ion is O^{5+}, whose formation from an O^{4+} ion requires the input of 114 eV of energy. This is eight times more energy than is needed to ionize a hydrogen atom. How are these highly charged ions produced? Ionization by a single photon can probably be ruled out on the grounds that there are hardly any stars in the Galaxy that produce significant numbers of 114 eV photons. If we make the alternative assumption that the ionization is maintained by collisions among the gas particles we can calculate that the temperature of the coronal gas is at least 3×10^5 K.

The second way of detecting coronal gas is by the X-rays it emits. The X-rays are a mixture of ionic emission lines and continuum radiation. Both kinds of radiation are the indirect result of the collisions between the gas particles. The collisions are like those which generate light and radio waves in an H^+ region, but because the coronal gas is so much hotter, the collisions are more violent and the resulting photons are more energetic. From the wavelengths and strength of the X-rays emitted by the coronal gas astronomers estimate that its temperature is about 10^6 K and its density is only about 0.003 ions cm^{-3}. Coronal gas is therefore 100 times hotter and at least 10 times thinner than any other kind of interstellar gas that we have met so far.

Coronal gas is hard to observe because it gives off almost no radiation that can be detected by ground-based telescopes. The few radio and visible-wave photons it does produce are swamped by the much stronger emissions from the warm ionized medium. Nevertheless, we now have enough ultraviolet and X-ray data from space to know that coronal gas fills large volumes of interstellar space. The obvious questions we would like answered are why is coronal gas so hot and where does it come from?

It turns out that the answers to these two questions are inseparable. The link arises because of the curious fact that when an ionized gas gets very hot it has great difficulty in radiating away its energy and cooling. To understand its difficulty we should recall how a gas usually loses energy;

collisions among gas particles cause atoms and ions to be excited into upper energy levels. They then descend spontaneously to their ground states emitting photons that carry energy away from the gas. In coronal gas this process is inefficient for two reasons. First, the density is so low that collisions occur infrequently. Second, at 10^6 K the most common atoms – hydrogen and helium – are stripped of all their electrons. With no upper energy levels that can be collisionally excited, these naked ions cannot contribute to cooling of the gas. The net result is that heat loss is extremely slow, and coronal gas can maintain a temperature of around 10^6 K for millions of years without any external source of energy.

Since hot gas naturally stays hot the problem is reduced to finding an origin for coronal gas. Fortunately, we do not have to look too far; supernova remnants appear to fit the bill admirably. As we discussed in Chapter 10, the expanding gases from a supernova explosion blast outwards into the surrounding interstellar medium with speeds of many thousands of kilometers a second. A large fraction of the energy of the supernova is transformed into kinetic energy of the gas as it roars outwards into space. When these expanding gases collide with stationary interstellar matter they push it outwards, leaving behind a comparatively empty cavity surrounding the now-dead star. This cavity contains hot, low density coronal gas. In the region of the collision shock waves are generated, and the gas's kinetic energy is converted to heat, raising the temperature to around 10^6 K. If the supernova is less than a million years old the hot gases will appear roughly spherical, as they do in the X-ray image of Tycho's supernova remnant (Figure 10.4). Even after a supernova remnant loses its shape and its identity, shock waves continue to move away from it at speeds of hundreds of kilometers a second. These shock waves continue to heat the interstellar medium and create more coronal gas. The interesting, and as yet unanswered, question is how much of the interstellar medium gets heated in this way?

11.4 Bubbles, tunnels, onions and sheets

An early assumption was that since supernovae are rare events, coronal gas should be confined to a few hot 'bubbles' surrounded by denser warm (8000 K) gas. The interstellar medium would thus have the structure somewhat resembling a Swiss cheese. Subsequent calculations, however, have suggested that these bubbles may continue to grow so large that they will eventually touch each other. The separate supernova remnants will then overlap and merge. The bubbles of coronal gas will coalesce with each other to form connected 'tunnels' that spread throughout the Galaxy. The interstellar medium would thus have a porous structure, with the coronal gas able to flow through the gaps in the warm gas, much as water flows through the holes in a sponge when it is squeezed.

If the 'tunnel' idea is correct it would imply that at least 50% of interstellar space is filled by coronal gas. Can it really be true that this thin hot gas, which was only detected in 1973, and which still cannot be seen by telescopes on the Earth's surface, fills over half the volume of our Galaxy? The question is controversial, but a number of astronomers think that the answer is yes. In the last few years there has been a radical change in the way that astronomers perceive the interstellar medium. The picture of a peaceful ocean of gas quietly filling the voids between the stars is being abandoned. In the new picture, space is filled with fast moving shock waves that continuously stir up the gas. Interstellar clouds are transient phenomena that are both generated and destroyed in the aftermath of supernova explosions.

How can this idea of a 'violent' interstellar medium be reconciled with the observations of neutral, ionized, and molecular gas we described in earlier chapters? First, we should note that even if coronal gas were to fill 90% of the *volume* of the Galaxy, its density is so low that it would still comprise only a tiny fraction of the total *mass* of the interstellar medium. Even in this new picture, the bulk of the interstellar gas is in a form easily seen by radio and millimeter-wave telescopes. What is considerably altered is our picture of where the neutral, molecular and photoionized gas is found. In earlier chapters we stressed that most regions of interstellar space are *either* predominantly molecular *or* predominantly atomic *or* predominantly photoionized. This idea is still valid, but it is now believed that the three types of gas can co-exist in the same cloud. Astronomers have calculated that a cloud of dense gas that is surrounded by coronal gas adopts a layered structure, something like an onion. The temperature is coldest and the density is highest in the middle. If the cloud contains enough dust to exclude ultraviolet radiation, molecular hydrogen will be found in its dark core. Surrounding the H_2 is the cool (80 K) neutral hydrogen that produces absorption lines like those seen in Figure 4.8. The cool neutral hydrogen is in turn surrounded by a layer of 8000 K warm gas, which may be either neutral or ionized, depending on how bright the local starlight is. Its outermost layer consists of gas which is in contact with the coronal gas which surrounds it. Heat is conducted inward through this layer from the hot surrounding coronal gas. It is gas in this outermost layer which produces the O^{5+} ultraviolet absorption lines which provided some of the first evidence for the existence of coronal gas.

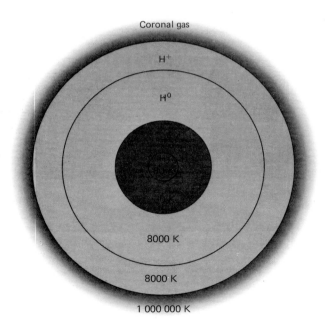

Figure 11.1. Theoretical interstellar cloud consisting of a number of onion-like layers of different phase gas. The coolest, densest gas is close to the center, while the outer layers are in contact with the hot coronal gas.

If these onion-like clouds are in contact with the million-degree coronal gas, why do they not all simply evaporate away? The answer is that there is a dynamic balance between the formation and destruction of interstellar clouds. Over a long enough period of time the numbers of clouds of different masses remain approximately constant. Individual clouds last typically for a few million years – a time that is long in the human scale, but very brief when compared to the 16 billion year age of the Galaxy. New clouds are formed when shock waves from supernovae sweep up fresh gas and compress it into denser clouds. The resulting small clouds may either grow into large ones by joining to each other when they meet, or they may shrivel away by evaporation. The clouds that grow continue to do so until conditions become ripe for the formation of new stars in their interiors – a subject we defer to the next chapter.

So far, we have said very little about the shapes of clouds. There is a strong temptation to depict them as spheres, as in Figure 11.1. The temptation arises because scientists are naturally inclined to attack simple problems before difficult ones. Because the sphere is the only perfectly symmetric three-dimensional shape, calculations of processes such as heat conduction are far simpler for spherical clouds than for odd-shaped ones. Astronomers therefore try to develop general ideas about cloud structure on the basis of hypothetical spherical clouds, then adapt these ideas if necessary and if possible to the more realistic shapes that we believe real interstellar clouds to possess. What are these shapes? Unfortunately they are very difficult to determine. Because our view of a particular cloud is always from the same angle it is almost impossible to know how thick a cloud is in the direction along our line of sight. This difficulty is acute in the cases of clouds which appear to us to be long and thin (e.g. Figure 11.2). Are we seeing a rope-

Figure 11.2. Real interstellar clouds rarely have simple shapes. It is often difficult to determine whether clouds such as this are rope-like, or sheet-like.

like cloud from the side, or a sheet-like cloud from the edge? In most cases we cannot tell, but it is strongly suspected that much of the cool gas in the Galaxy lies in thin sheets that mark intersections of shock waves from different supernovae.

The shapes of clouds may also be affected by several other factors not yet mentioned in this chapter. Interstellar magnetic fields are certainly important in some cases. Elsewhere, the rotation of the cloud, and of the Galaxy itself, can produce forces that distort the cloud from its original shape. The gravitational pull of distant stars drags interstellar gas towards the plane of the Milky Way disk. Finally, if the cloud is large enough, its own internal gravitational forces will cause it to shrink under its own weight, setting the cloud on the road to its conversion into a new star.

11.5 The local neighborhood

Where do we, in the Solar System, fit into this picture? Are we in a cloud or a bubble? Can we draw a map of the gas and dust within a few hundred parsecs of the Sun? These questions are tough for astronomers to answer, because they require that we take a very broad view of the subject. The view is broad in the literal sense of encompassing 360° all around us, and broad in the technical sense of requiring a synthesis of many different kinds of data. One of the problems is that different observations refer not only to different interstellar materials, but also to different properties of these materials. For example the 21-cm emission line gives wonderfully detailed maps of the neutral gas but it cannot distinguish warm from cold gas. Ultraviolet absorption lines can tell us a great deal about the physical conditions in clouds between us and a star, but the data are only available in a few specific directions in the sky. X-ray maps can give us a temperature for coronal gas, but no direct information about its distance. The picture that astronomers have put together is therefore extremely patchy.

The first clue to our local environment is the fact that we see X-rays coming to us from all directions in the sky. It tells us that the Solar System is surrounded by coronal gas; if we had been located inside a substantial cloud the X-rays would have been absorbed by neutral hydrogen and helium and would have been much dimmer than they are. We can get an idea of how far our local patch of coronal gas extends by looking for signs of neutral or photoionized gas in the ultraviolet spectra of stars at different distances. We find that in some directions the coronal gas extends at least 200 pc from the Sun, while in other directions cold neutral gas is found within about 50 pc. The Sun thus appears to lie near the edge of a 'bubble' of coronal gas. The most likely origin for this bubble was a supernova explosion in our neighborhood some 100 000 years ago. Unfortunately there is little chance of our ever finding the neutron star or black hole remains of the star that exploded.

The local bubble is by no means free of neutral gas. Faint Lyman-α absorption lines can be seen in the ultraviolet spectra of some of our closest neighbor stars. They imply that we are sitting at the edge of a small cloudlet of neutral gas, about 5 pc in diameter. The density of this cloudlet is only 0.1 particles cm^{-3}, and its temperature is around 8000 K – values that are confirmed by the solar backscatter experiments which we will describe in Chapter 13. The cloudlet is therefore comprised of warm neutral gas, not the much denser gas that is seen in the 80-K 'cool clouds'. When we look at the stars in the constellation of Sagittarius we are looking through the cloudlet, but it contains so little dust that we cannot see any perceptible reddening of these stars. Ultraviolet spectra of stars farther out in the bubble indicate that it contains a number of regions of cool neutral gas, but there are not enough data to determine the shapes or sizes of these clouds. For

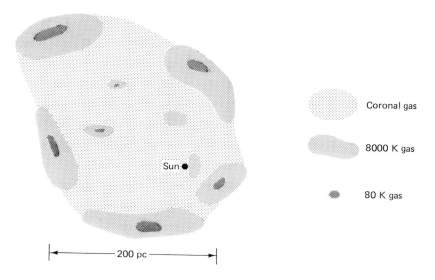

Figure 11.3. *The interstellar medium near the Sun. This highly schematic representation shows the Sun inside an irregular-shaped bubble of coronal gas. The bubble is surrounded by warm (8000 K) gas and by clouds. A few wispy clouds are found inside the bubble.*

want of a better theory we can guess that some of these cool regions resemble the 'onion' clouds we discussed in the last section.

The 21-cm hydrogen emission line is of little use for studying clouds within the local bubble, since it fails to distinguish between gas at different distances. Its value is in what it tells us about the more remote gas at and beyond the edges of the local bubble. Some of the loops and arcs that are seen in the 21-cm map of Figure 4.9 are almost certainly part of the shell of material swept up by the supernova that created the local bubble.

11.6 The galactic halo

We have alluded several times to the fact that most of the interstellar matter in our Galaxy is found in the plane of the Milky Way. The gas is confined to a thin layer because it is pulled there by gravity. The gravitational force that the gas feels is the sum of the pulls of millions of stars. The net effect of all these stars is that the interstellar gas falls to where the concentration of stars is the highest – the plane of the Milky Way. Interstellar gas reacts to the twin forces of gravity and pressure in much the same way that the Earth's atmosphere does; its density and pressure fall off gradually with height. As we discussed in Chapter 4, we can describe the thickness of an atmosphere in terms of its half-height, which is the distance upwards one has to go in order for the density to decrease by a factor two. Studies of the 21-cm line have shown that the half-height for the neutral hydrogen in the plane of the Galaxy is about 180 parsecs.

In our earlier discussions of half-height we ignored an important consideration – the gas temperature. Since pressure increases with temperature, a hot gas can support itself against gravity better than a cold gas. Consequently, the half-height of a hot gas is greater than that of a cool gas – a result that is justified mathematically in Appendix J. Since coronal gas is about 100 times hotter than the warmest neutral gas, the height of the coronal gas 'atmosphere' of the Galaxy is likely to be very much greater than the 180 parsec height of the neutral gas. It is therefore not surprising that coronal gas has been found well outside the plane of the Milky Way Galaxy. This outlying gas is referred to as the **halo** of the Galaxy. Its full extent is not known, but it extends many kiloparsecs into the regions in which few stars of any kind are found.

*Figure 11.4. The galactic halo. Because it is so hot, coronal gas can rise to great heights above the galactic plane. As it cools it recombines to form neutral clouds which fall back towards the plane of the Galaxy. Some of these so-called **high-velocity clouds** may be detected by their blueshifted 21-cm absorption lines.*

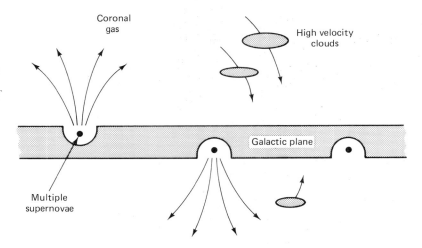

It is not only coronal gas which is found in the Galaxy's halo. Surveys of the sky at 21-cm wavelength have led to the discovery of a number of large atomic gas clouds in regions of the sky far away from the Milky Way. These clouds are referred to as **high-velocity clouds** because their Doppler shifts are much larger than would be expected from nearby gas in the galactic plane. Distances to high-velocity clouds are extremely hard to determine, but there is strong circumstantial evidence that some of them lie within a few kiloparsecs of the galactic plane. An intriguing feature of the high-velocity neutral clouds is that most of them have blueshifts, and must therefore be approaching the plane of the Galaxy. A possible explanation for some of the high-velocity clouds is that they are part of a **galactic fountain**. The fountain works as follows. Coronal gas is generated in the plane of the galaxy by supernovae; because of its long scale height it rises out of the plane into the halo. There it slowly cools and recombines to form neutral gas which, being cooler and denser than the coronal gas, sinks back to the plane of the Milky Way, becoming visible as the blueshifted high-velocity clouds. In this picture the coronal and the high-velocity neutral gas are therefore all part of the same slow cycle of the interstellar medium. We will discuss other possible origins for the high-velocity clouds in Chapter 15.

12 The formation of stars

In the last two chapters we have described the origin of interstellar matter and the various ways it can spend its life. To round off the story of the gas and dust in our Galaxy we must describe how its life cycle may be ended by its transformation into a new star.

The basic process that yields a new star is not in doubt. It is the collapse of an interstellar cloud under the strain of its own internal gravitational forces. Although the principle is simple, the details are appallingly complicated, and there are wide gaps in our understanding of when, where and how new stars and their attendant planets are formed. There are several problems. First, the birth of a star is very slow on a human time scale so that what astronomers observe are a number of different clouds in different stages of evolution. They must use experience to work out a suitable sequence of events for the phenomena they see, much as archaeologists deduce how species evolve by placing fossils in their correct sequence. Second, stars are formed inside dark clouds, severely limiting our ability to watch the process. A third problem is that as a cloud contracts to form a star it soon becomes so small that our telescopes can no longer pick out any details.

The fourth problem is the enormous mathematical complexity of the theories that are needed to describe the conversion of a gas cloud into a star. Some idea of the problem may be appreciated by comparing it to that of predicting the weather on the Earth. In both cases we are dealing with the motions of bodies of gas under the influence of forces like gravity and pressure and rotation. Indeed, many of the physical equations used by meteorologists are the same as those used by astrophysicists. Even though meteorologists have access to some of the most powerful computers in the world, erroneous weather predictions still occur. The far greater difficulty of astrophysical 'weather' predictions is obvious when we note that whereas the Earth's weather involves surface pressure variations of no more than a few percent, the transformation of an interstellar cloud to a star involves an increase of density of at least 10^{20}.

Despite the difficulties of the problem, many astronomers spend their professional lives trying to improve our understanding of how stars are born. Part of the reason for all this activity is that the birth of stars is one of the pivotal problems in astronomy. If it were not for stars forming now there would be no nebulae, no bright spiral galaxies, and probably no quasars. Without the birth of stars in the past, there would be no planets, no galaxies, and no us. The Universe would be totally dark.

In this chapter we will approach the problem of star formation from both the observational and theoretical points of view. First we must learn how to recognize the places in our Galaxy where new stars are currently being born.

12.1 Recognizing young stars

As explained in Chapter 5, the energy which a main-sequence star radiates into space is generated by the conversion of hydrogen into helium at its center. By comparing the rate at which energy is being emitted to the mass of hydrogen fuel in such a star we can estimate its potential lifetime. As we discussed in Chapter 5, the main-sequence lifetime of a star depends strongly on its mass; low mass stars are small, cool and long-lived: high mass stars, such as OB stars, are large, hot and short-lived. Our Sun is roughly halfway through its total main-sequence lifetime of 10^{10} years, but an OB star with a mass 40 times greater than the Sun would live for only a few million years. The fact that such bright stars are seen to exist today implies that star formation must have taken place within the last few million years. Since our Galaxy is *thousands* of million years old, it is reasonable to assume that somewhere in the Galaxy the same process is taking place even now. Recent estimates are that the equivalent of several stars the size of the Sun are born somewhere in the Galaxy each year. At the time the Galaxy first formed, however, the rate of star formation was many times higher than this.

Stars cannot in general move very far across the Galaxy in a few million years. Any OB stars we find must therefore still be fairly close to the places where they were formed. Searches for OB stars have shown that they are almost always found in the close vicinity of gas and dust clouds. We may therefore conclude that it is out of such clouds that new stars condense. These surveys have also shown that OB stars are almost always found in **associations** or **clusters**, containing hundreds or thousands of stars, and that most of these associations and clusters are in the spiral arms of the Galaxy. Examples of H^+ regions which are associated with clusters of OB stars include the Eagle Nebula (Figure 5.2) and the Orion Nebula (Figure 5.3).

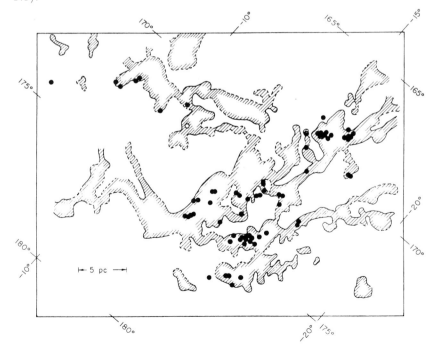

Figure 12.1. Sketch of parts of the constellations of Taurus and Auriga showing how the T Tauri stars (black spots) are usually found near the edges of dark clouds (shaded regions). The T Tauri stars drew attention to themselves by the emission lines in their spectra. They show several other characteristics that distinguish them from main-sequence stars and indicate that they have been formed comparatively recently.

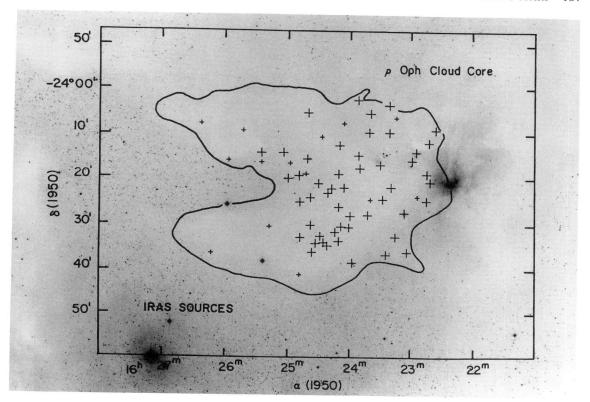

Figure 12.2. Negative image of the ρ Ophiuchi dark cloud. The crosses show the positions of infrared sources detected by the IRAS survey. Most of these infrared sources coincide with dense cores in this molecular cloud. Some are believed to be hidden T Tauri stars; others may be protostars. The ρ Ophiuchi dark cloud is also visible to the upper left of Figure 7.3.

Stars do not have to have large masses to be recognized as young. It is also possible to pick out youthful stars which have a mass about the same as that of the Sun. The signs are more subtle, and these stars are recognized primarily from unusual features in their spectra. One category of stars which are known to be young are the **T Tauri** stars – named after a particularly well-studied example of the type in the constellation Taurus. T Tauri stars are most often found near the edges of dark clouds, and distinguish themselves from normal main-sequence stars in several ways; their spectra show strong emission lines of hydrogen and other elements: they have variable brightness: they have larger diameters and a higher abundance of lithium than normal stars, and they sometimes emit more X-ray and infrared radiation than a regular star. A high concentration of lithium is regarded as a sign of youth because nuclear reactions inside a main-sequence star tend to destroy lithium over a period of time. Explanations for some of the other unusual properties of T Tauri stars will have to wait until later in this chapter.

T Tauri stars are recognized by their visible spectra, but even younger stars can be found by using infrared telescopes to search the interiors of interstellar clouds. The best places to look turn out to be the dense cores of certain molecular clouds. Dense cores are pockets of gas within a molecular cloud that have much higher than normal gas density. While some dense cores are cool, others emit strong infrared radiation – presumably from heated dust. Infrared emission from a dense core is taken as a sign that it contains either a newly-formed star or an object that is part of the way to becoming a star. Our understanding of these infrared sources is hampered by our inability to detect any visible light from them. Nevertheless, as we shall see, some of these infrared sources offer the potential of observing objects that are considerably younger than T Tauri stars.

A crucial feature of T Tauri stars and cloud-embedded infrared sources is that, like OB stars, they exist in groups, and in the vicinity of substantial clouds of interstellar matter. It is now generally accepted that all stars are formed by the gravitational collapse of interstellar clouds, and that stars form in groups rather than singly.

12.2 Gravity and clouds

While it may be true that all stars form out of interstellar clouds, it is certainly not true that all interstellar clouds form stars. In this section we will look at why some clouds foster new stars, but others do not.

The first question we ought to ask is what phase of interstellar gas do stars form out of? Unfortunately we cannot tell simply by looking at the gas near new stars. The gas we see now is very likely to have been ionized, heated or shifted around by the newly-formed stars. A simple physical argument, however, indicates that it is only in cool, dense molecular clouds that new stars can form. To understand why this is so we must look in more detail at how the force of gravity acts on interstellar gas.

Isaac Newton, in the seventeenth century, was the first to recognize that gravity is a universal force that causes every object to be attracted toward every other object. The strength of the force is described by Newton's law of gravitation. It states that the force depends only on the masses of the objects (essentially the amount of matter they contain) and on their separation. The force is stronger when the objects have large masses – which is why big things are generally heavier than small things – and is stronger when the objects are closer together. Mathematically, the strength of the force depends on the 'inverse square' of the distance between the objects. This means that if the separation of two objects is halved the attraction between them becomes four times as strong. If the separation increases by a factor 10, the force drops by a factor 100.

In principle, one could calculate the total force acting on an atom by summing the forces from all the other atoms in the Universe. Given that our Galaxy contains some 10^{68} atoms such a calculation is totally impractical. Fortunately, there are mathematical tricks that allow us to group large numbers of atoms together for the purposes of calculating gravitational forces. When we do this we find that in many places in the Universe almost all the gravitational force is produced by only one or two specific very large groups of atoms. At or near the surface of the Earth, for example, it is the group of atoms that make up the Earth itself (some 10^{50} of them) which produce almost all the gravitational force we feel. The force that pulls us toward the center of the Earth – our weight – far exceeds that which we feel from smaller objects (such as other people) nearby, or larger objects (such as the Sun) farther away. If we were to move away from the Earth in a spaceship our weight would decrease until at a distance of a few hundred thousand kilometers we would find that our attraction toward the Sun exceeded that toward the Earth. We would either fall into the Sun, or would orbit it indefinitely, like a planet. The Sun's gravity is, of course, the major influence on the motion of the planets in the Solar System, as well as on the matter within the Sun itself.

Outside the Solar System, in interstellar space, the strongest gravitational force is usually that toward the center of the Milky Way Galaxy. This force does not come from a specific object at the center itself, but is a kind of average force arising from all the stars and other matter distributed around the Galaxy. It is this centrally-directed gravitational force that holds the Galaxy together and maintains the stars in their orbits around its center. The force toward the center of the Galaxy is not always the dominant force however. In regions of the Galaxy where stars are clustered very close together, the net gravitational force may be directed into the star cluster itself. When this happens the cluster may maintain itself indefinitely as a recognizable unit, bound together by its own gravitational forces; any interstellar matter in such a cluster is also bound to it gravitationally. Another possibility, which is the one most relevant to the formation of stars, occurs only in dense interstellar clouds. Under the right conditions the dominant gravitational force may be that between the gas particles themselves, leading to a force that tends to pull the cloud in on itself. Remembering Newton's law of gravity, we can see that for an atom in the cloud to feel a strong force toward its center the cloud should have a large mass but a small size. It is thus in *dense* interstellar clouds that we might expect new stars to form.

12.3 Stability and collapse

Gravity alone does not determine if a cloud will become a star. What matters most is the balance between gravity and pressure; in a cloud, as in a star, there is a battle between gravity pulling gas inward and pressure pushing it out. Because the pressure in a cloud increases with gas temperature, the combination necessary for a cloud to collapse is *low* temperature and *high* density. These are just the conditions found in the cores of molecular clouds, and there is very little doubt that it is in this phase of the interstellar medium that new stars form. The fact that cool clouds are the sources of stars is at first surprising, given that stars are clearly much hotter than interstellar gas. As we shall see in the next section, however, the collapse of the cloud generates more than enough energy to heat the cool gas to the kinds of temperatures found in stars.

When we look more carefully at the battle between gravity and pressure in a cloud we soon notice an interesting result. The gas pressure at any point in the cloud depends only on the temperature and density of the gas at that point; it is therefore much the same for a small cloud as a large cloud. The gravitational forces in a cloud, however, depend strongly on the *total* mass of the cloud. A large cloud experiences greater gravitational forces and is therefore more likely to collapse than a small one, assuming that they have the same temperature and density. It can be shown that for a given temperature and density only clouds with more than a certain mass can collapse under their own gravity. This condition is called the **Jeans criterion** after the English astronomer who first formulated it (see Appendix M). For example, at a temperature of 10 K and a density of 100 molecules cm^{-3} a cloud would have to have a mass of over 20 M$_\odot$ in order to collapse. If the temperature of the gas were raised to 100 K, 600 M$_\odot$ would be needed in order for gravity to overwhelm gas pressure. Another way of looking at the Jeans criterion is to ask how dense a given clump of gas must be before it collapses under its own weight. For a 1-M$_\odot$ clump of gas at 10 K – a typical molecular cloud temperature – we find that the gas density must exceed 36 000 cm^{-3}. At this density the clump would have a diameter of 0.1 pc. It is from clumps such as these that we might expect stars like the Sun to form. Note that if a clump of gas is massive enough to start collapsing it has a good chance of continuing to do so, because its gravitational forces get stronger and stronger as it becomes smaller. For this reason we often talk of a gas cloud as either stable or unstable. Clouds with masses smaller than the mass given by the Jeans criterion are stable; clouds with masses above it are unstable and are liable to collapse.

The Jeans criterion gives us a tool for testing whether clouds are stable. The answer is surprising: when we survey the molecular clouds in our Galaxy we find that gravity is almost universally strong enough to overcome gas pressure. This result is true both for the giant molecular clouds as a whole, and for many of the smaller clumps, cloudlets and cores they contain. At first sight this result would suggest that stars are forming all over the Galaxy, but astronomers are reluctant to accept this idea. If all the clouds whose gravitational pressures exceeded their gas pressures were forming stars right now, a substantial fraction of the interstellar medium would be lost within a few million years. The loss of so much gas so quickly would have a devastating long-term effect on the appearance of our Galaxy.

Astronomers are generally reluctant to believe that such dramatic changes could be taking place simultaneously right now all over the Milky Way. They would prefer to believe that there is some influence at work inside the molecular clouds that inhibits star formation. Identifying the nature of this influence has proved difficult, but it now looks as if magnetism plays the major role in moderating the formation of stars.

12.4 Magnetic pressure

To understand how magnetism can support a cloud against gravity we need to reintroduce the idea of magnetic field lines. One of the reasons why physicists like to sketch the patterns of magnetic fields is that these sketches make it easy to visualize the direction of a magnetic force; the force always acts in such a way as to either shorten the length of a field line, or spread the field lines away from each other. In Chapter 9 we discussed the idea of flux freezing, and showed that if a magnetized cloud is forced into a smaller volume the magnetic field lines are pushed closer together. As illustrated by Figure 12.3, the magnetic field under these circumstances will generate

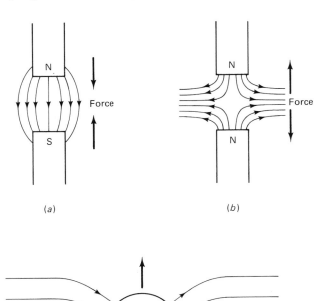

Figure 12.3. Magnetic forces act either to shorten field lines, as in (a), or spread them apart, as in (b). Magnetic fields in collapsing interstellar clouds (c) produce an outward force that is sometimes referred as magnetic pressure.

a force that opposes the collapse of the cloud. This is how magnetism can inhibit the formation of stars. It can be shown that there is a stability criterion for magnetically-supported clouds that is an analog to the Jeans criterion; for a given density and magnetic field, clouds with a mass smaller than some value are stable, clouds with a mass above this limit may collapse. Another way of understanding magnetic support is by noticing that magnetic fields behave in some ways like a gas under pressure; regions of high density (or field strength) attempt to spread out into the regions of low density (or field strength). The magnetic field in the compressed region exerts an outward force on the gas, or at least on those particles within it which are charged. We call this a **magnetic pressure.**

Magnetism may help explain another curious property of molecular clouds, which is that the Doppler broadening of their CO emission lines is much larger than expected. As we discussed in Chapter 4, the width of an emission line can tell us the average speed of the gas molecules in a cloud. When we apply this idea to the CO emission lines of molecular clouds we find that the gas motions are much larger than the thermal velocities characteristic of a gas at 10 – 50 K. Something inside the clouds must be stirring up the gas with random turbulent motions of many kilometers per second. What causes these motions? There are several plausible origins for the kinetic energy to get the gases moving. They include stellar winds from nearby young stars, ionization fronts, supernova explosions, and collisions between clouds. The difficulty is in distributing this energy around the cloud. The simplest method, using interstellar shock waves, does not work because shock waves quickly generate heat and fade away. A more promising method is to make use of **magneto-hydrodynamic waves**, or **Alfvén waves**. The easiest way to understand Alfvén waves is to use the analogy between gas pressure and magnetic pressure. Just as a sound wave is a moving disturbance of gas pressure, an Alfvén wave is a moving disturbance of magnetic pressure. Gas motions are spread through the cloud by wavelike oscillations of the magnetic field rather than through collisions between molecules.

Now we are faced with the opposite problem. If magnetic pressure is strong enough to support a cloud against its own gravitational collapse, how do new stars get formed at all? One possibility is to invoke some external influence that temporarily squeezes a cloud to a density that is high enough for instability to set in. The external factor could be a shock wave from a supernova or from a cloud–cloud collision. Another possibility is a process called **ambipolar diffusion.** Strictly speaking, magnetic pressure acts only on the charged particles in the gas – namely the electrons and the ions – which comprise but a tiny minority of the particles in a molecular cloud. The magnetic pressure is transmitted to the neutral atoms and molecules via collisions between the ions and the neutrals. Efficient though this process is, over a period of time some separation of neutrals and ions may occur; the neutrals may be pulled by a non-magnetic force such as gravity, leaving behind the charged particles. The process is a little like the sedimentation of solid particles in a cloudy liquid. Slowly but surely, dense cores of neutral gas form at various locations in the molecular cloud. At the places where the neutral gas collects, the gravitational forces increase and more gas is pulled toward the clump. So long as the gravitational forces are strong enough to overcome the gas pressure, neutral material will continue to diffuse toward the regions of high density. Eventually the clump becomes unstable as gravity becomes so dominant that the accumulation of material is unstoppable. When this happens the formation of a new star is assured. Astronomers sometimes refer to a clump of gas which is destined to become a star as a **protostar**.

*Figure 12.4. In a partially ionized cloud, neutral particles can become separated from charged ones, and will congregate where gravity is strongest. There is a frictional drag between the atoms and ions that limits the rate at which dense cores can form. This process is called **ambipolar diffusion**.*

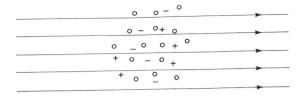

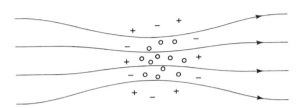

12.5 Protostars

The evolution of a protostar from an isolated fragment of a molecular cloud to a mature star takes hundreds of thousands, or even millions of years; astronomers attempting to follow this process must therefore resort to calculations rather than direct observations. The calculations involve estimating simultaneously the density, temperature, velocity, and heat flow in every part of a continually-changing ball of gas. The calculations rapidly become extremely complicated and can only be solved by computer. The astronomer writes a program and then gives the computer a description of the size, the shape and the temperature of the initial cloudlet. The computer program then applies the appropriate laws of physics and calculates what the protostar would look like at subsequent times. This kind of calculation is often called a **computer model**. Different models are 'constructed' by changing the initial conditions given to the computer or by applying slightly different theories of how the gases in the protostar interact with each other. The predictions of these computer models can then be compared with what is seen with telescopes.

Because of the complexity of the calculations, astronomers cannot always be entirely realistic in the way they set up their equations; it is very difficult to write programs that can take account of all the different shapes that cloud fragments could have. There is also the problem that there is no real beginning to the process of star formation; gravity, pressure and magnetism have been shaping the gas since the beginning of the Galaxy, so an astronomer has to make a somewhat arbitrary choice of where and how to start the model. Fortunately the evolution of a protostar does not depend too much on the shape of the original cloud. The reason lies in the way that a protostar collapses. It does not simply deflate gradually like a balloon. What happens is that as matter starts to fall inwards, a core of high density gas quickly forms, made up of material that has already reached the center. The subsequent motions of the outer parts of the protostar depend much more on what is happening at the center than on what is happening around the edges. The gas picks up kinetic energy as it falls toward the center; when it reaches the core this kinetic energy gets converted to heat by means of shock waves around the surface of the core. At first this heat is radiated away into space, but as the core grows to about 0.01 $M_\odot$ the heat can no longer get radiated away quickly enough and the growing core of gas becomes hotter and hotter as well as larger and larger. The core is the part of the protostar which will become the new star.

As a protostar evolves, the heat that is building up in its core tries to escape. The heat cannot be radiated away directly because the core is surrounded by the thick dusty outer layers of the collapsing protostar. What happens is that the dust absorbs the heat, gets warmed by it, and then radiates it into space as infrared energy. The bulk of the energy of a protostar is emitted between wavelengths of 3 and 300 μm (Figure 12.5). Infrared protostars provide us with the first glance we have of newly-formed stars, but the view is often a frustrating one. What we would like to see are visible and ultraviolet photons from the core itself, which could give us information about gas temperatures, densities, and motions surrounding the newly-forming star itself. Unfortunately, all we actually see at infrared wavelengths are the outer layers of warm dust which have been heated from within.

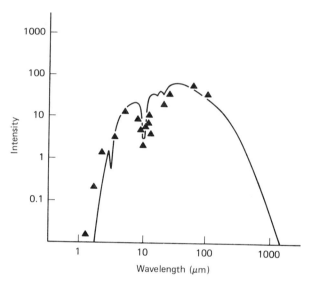

Figure 12.5. Protostars emit most of their energy at infrared wavelengths. This object currently radiates 2 $L_\odot$ of infrared power and will probably end up as a star of about the same mass as the Sun. The triangles are observations, and the lines are theoretical calculations of the expected emission from a protostar. Note the dips at around 3.1 μm and 9.7 μm which are absorptions by silicate and ice particles.

If the total amount of matter in the collapsing protostar is not too large then all of it will fall onto the core. For a protostar with the mass of the Sun this takes something like 10^5 years. However, the resulting star is still too large and too cool for the thermonuclear conversion of hydrogen to helium in its center. It is therefore termed a **pre-main-sequence** star. Over the next 10^7 years it slowly radiates away much of the heat it acquired during its infall stage. As it radiates this excess heat into space it contracts and gets hotter. Eventually nuclear reactions start and the star passes into maturity. T Tauri stars are believed to be in the pre-main-sequence stage of their evolution.

The larger the mass of a star the shorter is its pre-main-sequence phase. In the extreme case of a protostar core that has accumulated more than about 7 $M_\odot$, contraction to the point of thermonuclear energy generation takes place while the core is still growing. The resultant star has no true pre-main-sequence phase. A soon as the infall ceases the star is ready as a main-sequence star. In fact, it may be the star itself that cuts off the accretion, either by ionizing the infalling gas or by blowing it away via radiation pressure or a stellar wind.

12.6 Tools of the trade: computers in astronomy

Most professional astronomers spend far more time looking at a computer screen than through a telescope. Computers of all kinds, from humble desk-top PCs to ultrafast supercomputers, are indispensable to modern research. Astronomers make use of computers in four main ways.

First, they are used to control telescopes and the instruments that are attached to them. At a modern observatory an astronomer who wants to study a particular star simply types its position into a computer, which then moves the telescope to the appropriate angle in the sky, taking into account such factors as the time of day, precession, refraction, and even irregularities in the telescope's bearings. The same computer is often used to control the instruments on the telescope, changing filters, rotating diffraction gratings to the correct angle, timing exposures and storing the received data on computer tape. In some telescopes, particularly those operating at radio wavelengths or in satellites, the observations are totally automated, and the astronomer can literally go to bed while the data are being collected.

Second, astronomers use computers to analyze their data. Often, the information that an astronomer needs is hidden within a vast array of numbers collected at the telescope. A few kinds of data can be reduced by hand, but many observations would not even be attempted without the certainty that computers would be available to analyze the data. The VLA radio telescope (Figure 4.4), for example, produces data that is totally indecipherable until it has been manipulated thousands of times by a computer.

Third, astronomers use computers in the same way that members of many other professions do; they use word processors to write reports, books, and grant proposals, they communicate with each other by electronic mail, they use databases to keep track of the flow of information, and graphics programs to prepare their illustrations for publication. They also may have to balance the observatory's budget on a spreadsheet.

Fourth, computers are used by theoretical astronomers to perform complex calculations. The most difficult problems are those that involve watching how stars, clouds, or galaxies evolve over time. Even if the underlying physical principles are simple, only rarely can such problems be solved just by manipulating a few equations. Usually, several equations must be solved simultaneously for each of thousands of points in both time and space. The computers needed for the largest of these calculations can cost tens of millions of dollars, and are among the most powerful of all computers used for scientific research. They are the domain of specialists in computational physics and making use of one can be rather like visiting an observatory to watch a star. A theoretical astronomer might have to apply to a committee for permission to use a supercomputer for a few hours, just as an observational astronomer would apply to use a telescope to observe a galaxy for a night. The theoretician will set up the equations as a program for the computer to follow and watch what happens, in much the same way as the observer will gather data from a spectrometer.

Major advances in astronomy have usually come about either by making totally new kinds of observations, or by noticing connections between apparently disparate phenomena. Computers have helped enormously with the first kind of task; whether programs based on artificial intelligence can make contributions of the second kind remain to be seen.

12.7 The problem of rotation

In our discussion of star formation we have up till now ignored the possibility that the protostar might be rotating as it collapses. A simple physical argument shows that we have no right to do so, and that rotation probably plays a very significant role in determining how a star forms.

The problem is the law of conservation of angular momentum, which was first formulated by Isaac Newton. Angular momentum is a measure of the amount of rotation a body possesses. It depends on the mass, the size and the rotation speed. The law explains in detail how the rotation speed of a body has to increase if its size decreases. A simple demonstration of the law in operation is presented by an ice skater performing a spin. The skater starts off rotating on one skate with both arms and one leg outstretched. To spin faster the limbs are retracted to form the thinnest possible profile. The same effect happens in a protostar; if an interstellar cloud starts off rotating at all, it will rotate faster as it shrinks.

There is little doubt that molecular clouds, and hence protostars, do rotate. We often see slightly different Doppler shifts from opposite sides of a cloud, as though one side is moving away from us and one side toward us. Furthermore we know that the whole Galaxy is rotating about once every 2.4×10^8 years, and it would be very surprising if the molecular clouds were not rotating at least this fast. Ponderous though this rotation is, when it is amplified by the effect of angular momentum conservation its presence has profound effects on the evolution of a protostar. Consider a protostar that starts off with a diameter of 0.1 pc and a rotation period of once every 2.4×10^8 years. If this gas were to shrink to the size of the Sun while conserving angular momentum the resultant star would rotate once every 30 minutes! This rotation is far faster than we actually observe for any normal star; the Sun, for example, spins on its axis once every 28 days. In fact a star spinning at twice per hour could not even exist; the rotation would be so fast that the star would be spun out into a disk by centrifugal forces.

Clearly, if stars are to be born, some way must be found of getting rid of most of the angular momentum contained in the original protostar. There are at least three methods which seem to go some way to alleviating the problem.

First, there is the mechanism known as **magnetic braking**, a process that is probably most important during the early stages of the collapse of a cloud. Because of flux freezing a rotating cloud will twist the lines of magnetic field (Figure 12.6). The twisting leads to both stretching and squeezing of field lines which, as we discussed in Figure 12.3, results in a force that opposes the rotation. The magnetic field serves to transport angular momentum away from the dense core and into the outer parts of the cloud. Magnetic braking can shift the direction of a cloud's rotation, which may partly explain why interstellar clouds do not generally spin in the same direction as the Galaxy.

Second, a protostar may split into two pieces and collapse to form a **binary** pair of stars, rather than a single object. More than half the stars in the Galaxy have one or more partner stars which they orbit around. In a binary system much of the angular momentum of the original cloud is taken up by the orbital motion of the stars around each other. Unfortunately we do not yet have any satisfactory theory for what causes a rotating protostar to break up into a number of objects.

Third, a rotating disk of gas and dust may form around the protostar's core. Gas spiralling inward falls onto the disk if it has too much angular momentum to reach the core itself. A rotating disk of molecular gas has been detected around the young star HL Tauri by its millimeter-wave CO

146 THE FORMATION OF STARS

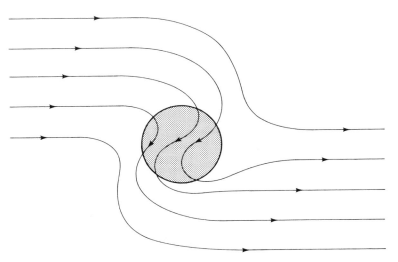

Figure 12.6. A protostar will spin faster as it collapses, pulling the magnetic field into a spiral as it does so. The stretched magnetic field lines produce a twisting force which will tend to slow the rotation of the protostar and transfer the angular momentum to the outer regions of the cloud. This process is called **magnetic braking**.

emission. Infrared emission from dust in a disk has been detected around several young stellar objects. The infrared emission arises because the disks are warmed both by the central star, and by the kinetic energy of infalling matter hitting the disk. Disks are thought to be the main cause of excess infrared emission from T Tauri stars, and it is possible that all young stars are surrounded by disks for part of their lives.

What happens to the disks? One possible fate is the formation of a system of planets around the star. Two interesting features of our Solar System are that all the planets revolve around the Sun in the same direction, and that there is much more angular momentum tied up in the orbits of the planets than in the Sun itself. There seems little room to doubt that our system of planets was formed out of a disk rotating around the young Sun. Unfortunately, it is technically extremely difficult to observe planets around other stars and most of the attempts to do so have had very limited success. The IRAS satellite, however, found evidence for a disk of heated dust around the star β Pictoris, which is a *not* a young star. Although the dust around β Pictoris does not constitute a planetary system in itself, it does establish that there is at least one middle-aged star beside the Sun which is surrounded by orbiting solid particles.

12.8 Outflows and winds

Despite the confidence with which astronomers expound their theories of star formation, *direct* evidence of the collapse of an interstellar cloud is in embarrassingly short supply. Astronomers have looked hard for Doppler shifts that would indicate inward motions in a protostar, but with a few exceptions they are not seen. One problem is that it is hard to find protostars in their earliest phase, because until a hot core is formed they give out little energy. A second problem is that the infall velocities for star formation are small, especially if they are limited by ambipolar diffusion. The third problem is that many protostars and young stellar objects show much stronger evidence for *outflow* of gas than for infall.

The first strong clue that the star formation process might lead to rapid outward movements of gas came with the discovery of Doppler shifts of over 100 km s^{-1} in certain H$_2$O maser sources. It was soon realized that the gravitational collapse of a protostar could not produce velocities this large even in the absence of magnetic pressure. Some other kind of force had to be accelerating the gas to these high speeds. Additional evidence for large

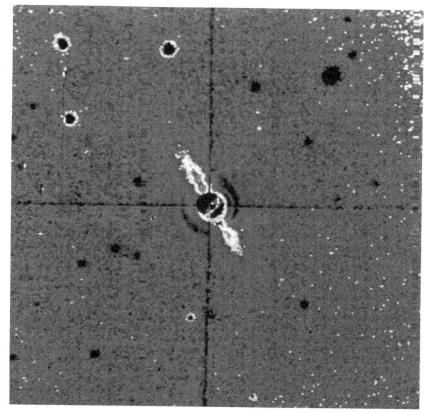

Figure 12.7. Image of the star β Pictoris, showing what many astronomers believe to be an edge-on disk of solid particles. In this picture most of the light from the star itself has been deliberately blocked out to cut down on dazzle. Although this picture was taken by visible light the disk was first noticed because of its strong infrared emission. The cross, and the dark rings around the central star are produced by the telescope.

velocities in the vicinities of new stars was then found in the Doppler shifts of millimeter-wave CO emission lines, and in the motions of certain small visible nebulae called **Herbig–Haro objects**. The most direct evidence for outflows, however, came with the discovery of **bipolar flows** around a number of protostars and young stellar objects.

Bipolar flows are streams of gas flowing in opposite directions from a newly formed star. The word bipolar is used because it is assumed that the flows are lined up along opposite (e.g. north and south) poles of the central star. Bipolar flows are most easily found by studying the Doppler shifts of millimeter-wave CO emission lines, but signs of bipolar flows are sometimes also seen at visible or infrared wavelengths (see Figure 8.3, for example). The stars from which they originate are sometimes visible, and sometimes detectable only by their infrared emission. Bipolar flows are always associated with protostars or young stellar objects; they are quite different from the bipolar nebulae which are formed around old stars, and which we discussed in Chapter 10. Most bipolar flows are situated deep inside molecular clouds.

There is controversy about what causes bipolar flows, but there is general agreement that rotation plays a crucial role. Some theories attribute the flow to the complicated combination of gravitational, magnetic and centrifugal forces that act on the gas as it falls inward toward a central disk. Under the right conditions these forces could cause some gas to spiral outward around the rotation axis of the protostar producing the appearance of a bipolar flow. The twisting outward motion of the gas serves the additional function of carrying away some of the angular momentum of the star. There are other theories that attribute the bipolar flow to a wind generated on the

Figure 12.8. A bipolar flow in the molecular cloud B335. Observations of millimeter-wave CO molecules reveal redshifted and blueshifted gas on oppposite sides of a hidden infrared source (shown as a small square). The infrared source radiates about 8 $L_\odot$ of power, and will probably end up as a star with a mass of no more than 2 $M_\odot$.

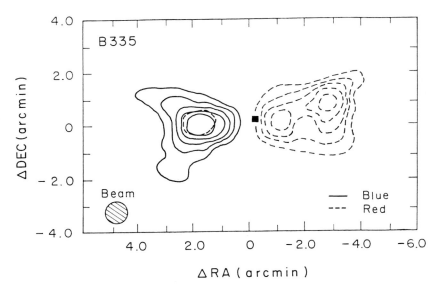

star itself. The wind is funneled along the rotation axis of the star because in a rotating protostar the density is thinner above the poles. In both theories, *infall* toward the disk (equivalent to the equator of the star) may be proceeding at the same time as *outflow* along the axes. Determining whether there is a net gain or loss from a young star is very difficult because the inflowing gas, being colder and moving inwards very slowly, is much more difficult to detect than the outflowing gas.

Many astronomers now believe that all stars may go though an outflow stage during their early lives. One reason for believing this is simply that bipolar flows are found in such a large percentage of molecular cloud cores. Other evidence for widespread outflows includes the observation of strong stellar winds from OB stars, from T Tauri stars and from other young stars. It is one of the paradoxes of observational astronomy that the birth of a star is recognized by the outflow rather than by the inflow of matter.

12.9 High mass stars

In this chapter we have been mainly concerned with the formation of stars which have masses about equal to that of the Sun. Although the births of high mass stars (more that about 10 $M_\odot$) are much rarer than the births of low mass stars they deserve special mention for several reasons.

First, there is some evidence that high mass stars and low mass stars are not always born in the same place. There are no signs of any new stars with masses more than 3 $M_\odot$ in the Taurus clouds, although the Orion region harbors new stars of all masses. One idea that is gaining ground is that star formation is **bimodal**. There are some molecular clouds which produce high mass stars, and some which produce only low mass stars. The difference may be in the clouds themselves, or in the way that star formation is triggered or cut off in different places. One suggestion is that low mass stars form out of relatively undisturbed molecular clouds, but that high mass stars need the extra impetus of a collision between molecular clouds to get them started. Because collisions are more likely where molecular clouds are packed closer together, high mass stars would be more concentrated in spiral arms than low mass stars, as observed.

Second, as we discussed in Section 12.5, there is no real pre-main-sequence stage for high mass stars, so their formation is extremely rapid. They reach

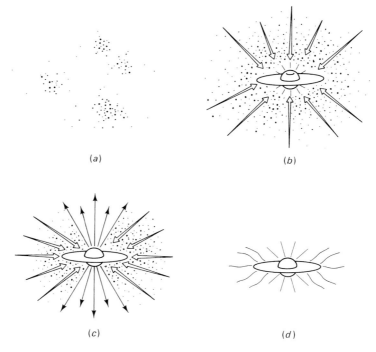

Figure 12.9. *The formation of a rotating star. First, cores form inside a molecular cloud. Second, a protostar forms, surrounded by a rotating disk. Third, a bipolar flow breaks out along the rotational axis of the protostar. Fourth, the pre-main-sequence star is left surrounded by a disk out of which planets may subsequently form.*

maturity while still embedded deep inside a molecular cloud. If the new main-sequence stars are OB stars they will form H^+ regions that are completely enveloped by molecular gas, as in W49 (Figure 5.7).

Third, places where new high mass stars are being formed are among the most prominent regions in the Galaxy. The reason is simply that OB stars and their precursors pour enormous amounts of power into regions of space that are very rich in interstellar matter. The energy absorbed by the gas and dust has to be reradiated back into space, with the result that the clouds and the nebulae become very easy to see at almost any wavelength. The H^+ regions emit light, ultraviolet and radio waves, the molecular gas emits millimeter-wave spectral lines, and the dust emits infrared radiation. The brightness of high mass star formation regions at infrared, radio and millimeter wavelengths allows them to be seen right across our Galaxy, unimpeded by obscuration. Molecular clouds and H^+ regions, together with the 21-cm atomic hydrogen line, have provided us with most of what we know about the structure of the distant parts of the Milky Way, such as the location of its spiral arms. In other galaxies, too, it is often the H^+ regions that display the locations of spiral arms most clearly.

Fourth, as we discussed in Chapter 10, high mass stars end their lives as supernovae. The effect of a supernova exploding in a star formation region must be devastating, though we have not had the chance to witness such an event in our lifetimes. Depending on its location, the blast could destroy a molecular cloud, or could shut off star formation completely over a substantial region of space. On the other hand, a number of near-simultaneous supernovae could be the trigger that sets off more star formation in a neighboring part of the Galaxy. Such an idea has been put forward to explain why star formation sometimes appears to have taken place as a sequence of events; the same idea has been suggested as one of the ways that spiral arms may form. One phenomenon related to supernovae that we certainly can see is the existence of **superbubbles** in the interstellar medium. These are huge bubbles of gas surrounding clusters of mature O and B stars. The bubbles result from the explosions of the fastest-living members of the clusters and from the effect of the enormous winds that blow from very young O and B stars. Several of these loops can be seen in Figure 4.9. The existence of these very large structures in the diffuse interstellar medium is another sign of the power that is released into the Galaxy when high mass stars form.

12.10 The fate of the interstellar medium

A major question that we have not yet addressed is the evolution and fate of the interstellar medium. We have discussed the competing processes of its regeneration from stellar winds and its depletion in the form of new stars, but we have not yet asked how these two processes balance. Is there an equilibrium, or is the total amount of interstellar matter subject to long-term change?

Unfortunately there is no single observation that will tell us how fast the interstellar medium is being used up. An inventory has to be taken of the stars and the gas in the galactic disk, together with estimates of the rate of formation of new stars and the total rate of mass loss from old stars. Ideally this inventory would cover the whole Galaxy, but in practice it has only been possible to survey the solar neighborhood within about 1 kpc of the Sun. Even within the solar neighborhood we have to take proper account of the fact that the stars and the interstellar gas are not uniformly mixed with each other. For one thing the layer of stars in the galactic plane is thicker than the layer of gas and dust. For another, star formation occurs only in a few dense clouds. To get a realistic average we have to imagine that all the contents of the Galaxy's disk is compressed into a very thin layer, in which we can itemize the various components in terms of their concentration per square parsec of disk.

When we perform this inventory we find that the total amount of mass in the form of stars is in the range 27–50 $M_\odot$ pc^{-2}. The uncertainty depends on whether one counts the stars directly or estimates the total amount of matter from measurements of the gravitational field of the Galaxy. Interstellar matter contributes about 12 $M_\odot$ pc^{-2}, thus making up some 15–30% of the total mass in the solar neighborhood.

New stars are formed in the solar neighborhood at the rate of 5×10^{-9} $M_\odot$ pc^{-2} yr^{-1}. The rate of return of gas into interstellar space by red giants, supernovae and the like is only about one fifth of the star formation rate, implying a net loss to the interstellar medium of about 4×10^{-9} $M_\odot$ pc^{-2} yr^{-1}. If this loss rate were to continue indefinitely all the interstellar gas in the solar neighborhood would be used up during the next 3 billion years. After this no further star formation would be possible.

To understand why the interstellar medium is disappearing at this rate we need to review a few important points about the lives of stars. As was mentioned in Chapter 5, the length of time that a star spends in its hydrogen-burning main-sequence stage depends on its mass. OB stars with masses of 10 $M_\odot$ or more have short lifetimes of a few million years. Our Sun has a lifetime of about 10 billion years, which is somewhat more than half the present age of the Galaxy. Stars smaller than about half the mass of the Sun are able to stay on the main sequence for times that are longer than the current age of the Universe. Since substantial mass loss from stars generally does not start until after stars have left the main sequence we must conclude that any interstellar matter that goes into making low mass stars is essentially locked up for good. Matter is recycled back into the interstellar medium only from high mass stars, which are greatly in the minority; only 10% of the total mass of the stars in the solar neighborhood is made up of stars with masses larger than the Sun.

The conclusion that our Galaxy will run out of interstellar matter in a few billion years must be treated with caution for several reasons. First, the statistics apply only to our local neighborhood; other parts of the Galaxy may behave differently. Second, the estimate of star formation rate depends critically on the statistics of faint stars, which are intrinsically difficult to study. Third, it is possible that new interstellar matter is entering the Galaxy from outside. In Chapter 11 we alluded to the high-velocity clouds of atomic hydrogen which lie well outside the galactic plane. Although the best explanation for most of these is that they contain gas which was earlier ejected from the plane of the Galaxy it is possible that some of the gas in the high-velocity clouds is reaching the Milky Way for the first time. If so, the rate of infall might be enough to balance the loss of interstellar matter to star formation, and the interstellar medium could have a future longer than 3 billion years.

13 The interplanetary medium

So far, this book has been almost exclusively concerned with the interstellar medium. The word 'space' in the title of this book, however, has a broader meaning. In this chapter we will take a look at the **interplanetary medium**. The phrase refers to the gaseous or dusty material we find within the confines of the Solar System. As we shall see, there are major differences between the interstellar and interplanetary media, both in the constituents themselves and in the forces and balances that maintain them. The most striking difference is that the gas and the dust within the Solar System have virtually nothing to do with each other. They are concentrated in different parts of the Solar System, move in different directions, have different origins, and are subject to quite different sets of forces. In this chapter we will also examine what can be said about the boundary between the interplanetary medium and interstellar space.

13.1 Interplanetary dust

We can see interplanetary dust with our naked eyes, so long as we know where and when to look. On a dark moonless evening, well away from city lights, a faint band of light may be seen stretching up from the western horizon shortly after twilight ends. It can also be seen in the east before dawn. The phenomenon is called the **zodiacal light**. It is caused by sunlight scattering off interplanetary dust particles as they orbit the Sun. Because the orbits of the dust particles lie mostly in the same plane as the orbits of the planets, the light glow is seen only in the parts of the sky through which the planets appear to move. Hence the name 'zodiacal light'. In some ways the zodiacal light can be thought of as a reflection nebula within our Solar System. One important difference, though, is that the color of the zodiacal light is almost exactly the same as that of the Sun; as we discussed in Chapter 7, the scattered light in interstellar reflection nebulae is bluer than the light which illuminates the dust grains; the reason for the color change in the interstellar case is that very small particles scatter blue light better than red light. The lack of color change in the interplanetary case implies that most, if not all, of the dust particles that cause the zodiacal light are significantly larger than the wavelength of light. Many of them must be more than 10 μm in diameter, making them much bigger than the typical interstellar dust grains we met in Chapter 7.

With the advent of the space age a new means of studying the zodiacal dust grains has become available. Like interstellar dust grains, zodiacal dust grains absorb as well as scatter light. Particles in the vicinity of the Earth

Figure 13.1. Interplanetary dust particles can be seen by the naked eye as the zodiacal light, a band of scattered sunlight stretching upwards from the horizon after sunset or before dawn on dark nights.

absorb enough energy to raise their temperatures to several hundred degrees kelvin, causing them to emit strong thermal infrared emission. So powerful is this emission that in the 10–40 μm range it swamps the radiation from the distant stars. The IRAS satellite, although designed primarily for surveying the infrared emission from our own and from other galaxies, was extremely sensitive to the zodiacal infrared emission and allowed astronomers to make more detailed maps of the location of interplanetary dust than had been possible using visible light observations.

Large interplanetary grains – those at least a millimeter in size – can be seen in another way. In their orbits around the Sun interplanetary dust particles run the continual risk of bumping into a planet such as the Earth. The effect of a collision between a dust grain and the Earth is devastating for the grain. As it plunges into the Earth's atmosphere with a speed of many kilometers per second, air friction raises its temperature to white heat, causing it to sublimate to a gas in a few seconds. The glow from a dying dust grain provides us with the magnificent spectacle of shooting stars, or **meteors**, which can be seen on most dark nights, given a little patience.

Not all interplanetary particles are destroyed by the Earth's atmosphere. If an object is large enough, only its outer parts will have time to melt during its journey through the atmosphere. This is how **meteorites** – lumps of interplanetary rock or metal weighing anything from a few grams to a few tons – are able to reach the Earth's surface. It is estimated that hundreds of tons of meteors and meteorites either reach the Earth's surface or burn up in its atmosphere every day.

Very small interplanetary grains are also capable of surviving entry into the Earth's atmosphere. Because they are so light they are slowed down in the thin upper layers of the Earth's atmosphere and sink gently through the thicker layers that would burn up larger objects. Small particles radiate away their heat better than large ones, so their temperatures do not rise as high. Particles that are 10 μm or more in size will be heated to less than 550 K – not enough to melt any minerals present. Once these small particles have been slowed down by air friction they float gently down towards the Earth's surface on a journey that will usually take somewhere between one and 60 days, depending on the particle size. For more than 100 years efforts have been made to collect small interplanetary particles. The earliest method used was to drag a magnet over the ocean floor to pick up iron-rich particles that might have collected there. Another method is to analyze grit that is found on top of bleak Arctic ice-fields. Such grit is likely to be far less contaminated by terrestrial minerals than are samples of material found on the rocky surface of the Earth. Since the 1970s, however, the best samples of interplanetary dust have been obtained from high-flying aircraft. Special instruments on a NASA U-2 aircraft scoop up microscopic particles which have recently entered the Earth's atmosphere from above. Particles collected in this way are between 1 and 50 μm in diameter. The advantage of using an aircraft over a satellite like the Space Shuttle is that the collection process is much gentler. In space a grain would hit a satellite at many thousands of kilometers per hour, and become seriously damaged as a result. The aircraft scoops them up at a speed of only a few hundred kilometers per hour.

Figure 13.2. An interplanetary particle which was captured by an airplane at high altitude.

When analyzed in the laboratory interplanetary dust particles are found to have a variety of shapes and compositions. Most of them contain many minute crystals made of minerals resembling those found in meteorites, including silicates and sulfides. The shapes of some of the crystals indicate that they condensed directly to a solid out of the gas phase. There are also carbon-based compounds and some minerals which have never been seen in natural samples before. The infrared spectra of the collected particles can be examined in the laboratory and compared with spectra obtained by astronomers observing interstellar dust in distant parts of the Galaxy. Quite strong similarities are found between the laboratory spectra and the spectra of interstellar dust grains found in the vicinity of newly-forming stars in molecular clouds.

13.2 The origin and fate of interplanetary dust

Dust grains within the Solar System have a much more eventful life than their counterparts in the interstellar medium. The most disruptive force they have to deal with is the radiation pressure of sunlight. The pressure of the photons flowing outward from the Sun causes small grains to be blown out of the Solar System. Paradoxically, radiation pressure acts in an opposite fashion on larger grains, causing them to be slowed down in their orbits and eventually to fall into the Sun. Interplanetary dust grains are also destroyed by collisions with each other, by being swept up by planets and by being vaporized by the Sun. Calculations indicate that the average dust grain in the vicinity of the Earth can survive for only about 10^5 years in the Solar System. Since this time is far shorter than the estimated 4.6 billion year age of the Solar System the dust must be getting replenished somehow. It has been estimated that an average of 80 tons of new dust must be produced each second to replace what is being destroyed.

Where does all this dust come from? Several possibilities can be eliminated fairly easily. It cannot all be coming from outside the Solar System, since the dust density in the local interstellar medium is much too low to provide the necessary 80 tons per second. The Sun is not a promising source either; there is no dust in the Sun itself and the density of the gas in its wind is too low for even simple molecules, yet alone solid particles, to form. The most likely sources are comets and asteroids.

Figure 13.3. Comet West in 1976. The tail of a comet is visible because of sunlight scattering off dust grains that have been released from the nucleus. Comets are a major source of interplanetary dust grains.

The nuclei of **comets** are small conglomerations of dust grains, rocky particles, water ice and other frozen gases. They normally orbit the Sun far beyond the planets Pluto and Neptune. They were formed at about the time the Sun and planets came into being, some 4.6 billion years ago, and have been cruising through the outer reaches of the Solar System since then, practically unaltered since the time they were formed. Comet nuclei are only a few kilometers in diameter and are normally much too faint to be seen, even with the largest telescope. Occasionally, however, a comet gets perturbed out of its distant orbit and is pulled towards the inner Solar System by the gravitational force of the Sun. When this happens the comet nucleus starts to warm up, as a result of the increased heat from the Sun. As the frozen gases in the comet evaporate, the dust particles trapped in the ice are released into space. The radiation pressure of sunlight blows these dust particles away from the nucleus, forming the long tail that is the most spectacular characteristic of a comet when it is near the Sun. Most of the light which we see from the tail of a comet is, in fact, sunlight scattered off dust grains, just as in a reflection nebula or in the zodiacal light.

Comets are not the only source of interplanetary dust. The IRAS measurements showed that some interplanetary dust particles have orbits that are almost identical to those of certain **asteroids**. Most of the known asteroids have diameters of a few tens of kilometers, though innumerable faint smaller ones must certainly also exist. Asteroids are generally believed to be primitive Solar System objects which for some reason failed to amalgamate to form a proper planet. There are enough asteroids in the Solar System that collisions between them must occasionally occur even now. Such collisions would pulverize the rocks and eject substantial quantities of dust into the Solar System.

While there is little doubt that comets and asteroids are the major sources of interplanetary dust, we can still ask whether it is possible that there are any genuine *interstellar* dust grains inside our Solar System. As yet, none has been convincingly recognized among the material collected by high-altitude aircraft. Recently, however, possible samples of interstellar dust have been discovered by quite another method. Minute crystals of silicon carbide have recently been found inside a meteorite which landed on the surface of the Earth. To understand the significance of this discovery we must recall that most stars contain more oxygen atoms than carbon atoms. As we discussed in Chapter 10, when dust forms in the winds of these stars, the silicon atoms combine preferentially with oxygen atoms to form silicate minerals. If the gas contains more carbon atoms than oxygen atoms, however, silicon carbide is formed instead of silicates. Our Sun and the planets definitely contain more oxygen than carbon; the same must have been true of the interstellar cloud out of which the Sun formed. As far as we can tell, at no stage in the history of the formation of the Solar System were there ever any substantial amounts of carbon-rich gas. Therefore, the silicon carbide crystals discovered in the meteorite must have been formed in the wind of a carbon-rich red giant star *before* the Solar System was formed. The grains managed to survive the traumas of the formation of the Solar System and become incorporated into the meteorite about 4.6 million years ago. The silicon carbide crystals can be recognized as once being interstellar because of their unusual composition. As analytic techniques improve it may be possible to show that some of the silicate crystals found within the Solar System were once also interstellar dust grains.

13.3 The solar wind

There is very little gas associated with the zodiacal dust particles. They are thus in a very different environment from their interstellar cousins which are usually associated with 200 times their own mass in the form of hydrogen and helium gas. The reason for the absence of 'zodiacal' gas is that comets and asteroids themselves have far less hydrogen and helium than one finds in the Sun or in the interstellar medium. As happened on the Earth and on other planets with low gravity, these light gases long ago drifted into space

Most of the gas in the Solar System is in the form of a wind of charged particles rushing outward from the Sun. In fact, to many astronomers, the phrases 'solar wind' and 'interplanetary medium' are interchangeable. The **solar wind** cannot be seen with the eye, even with telescopes, but indirect evidence for its existence started to accumulate in the first half of this century. Fluctuations in the Earth's magnetic field and in the brightness of the Aurora Borealis or 'Northern Lights' were traced to the injection of streams of charged particles into the Earth's upper atmosphere. These geomagnetic disturbances were found to occur a few days after major storms had occurred on the Sun's surface, so it was guessed that the particles came from the Sun. The existence of a solar wind of charged particles was also suggested by the motions of ionized gases in comet tails; they appeared to be accelerated by a force stronger than radiation pressure alone. The clinching evidence for the existence of a steady solar wind had to wait for the space age. Experiments on board unmanned spacecraft in the early 1960s revealed an ionized gas with a temperature of 100 000 K rushing away from the Sun at speeds in the range 300–800 km s^{-1} – far faster than anything else that moves in interplanetary space. The density of the wind is about 10 particles cm^{-3} near the Sun, decreasing as it expands into the outer reaches of the Solar System. Most of the particles are protons, helium nuclei and electrons, though there are a few heavier element ions as well.

We do not fully understand how the solar wind is generated, but we know it starts in the **corona** of the Sun. The corona is the uppermost layer of the Sun's atmosphere. It has a temperature of 10^6 K – far hotter than the 5800 K of the Sun's visible surface. Motions of gas in the corona are controlled almost entirely by the complicated magnetic field of the Sun, and it is likely that these fields play a major role in propelling the gas away from the Sun. The solar wind is ionized when it leaves the Sun, and stays

Figure 13.4. The solar corona is the outermost layer of the Sun's atmosphere. It is best seen during a total solar eclipse, as here, when the light from the Sun's bright disk is obscured. The corona is the source of the solar wind.

ionized all the way out to the edge of the Solar System; because the density is so low, ions and electrons in the solar wind have precious few opportunities to meet and recombine in the several years it takes to traverse the Solar System.

The solar wind is not at all steady. Changes in the condition of the Sun can lead to tenfold temporary increases in the wind strength, and cause the interplanetary medium to be extremely inhomogeneous. These fluctuations in the solar wind can be observed by radio astronomers; radio waves from distant objects such as quasars are bent and distorted as they travel through the interplanetary medium; these distortions show themselves as a kind of 'twinkling' or scintillation of the apparent strength of the radio signal.

13.4 Interstellar gas within the Solar System

The solar wind is not the only gas in the Solar System. Ultraviolet telescopes on spacecraft and rockets have detected the presence of a gentle wind of neutral interstellar gas flowing through the Solar System.

To understand how this interstellar gas was detected we should recall how ultraviolet absorption lines are produced in the interstellar medium. The strongest interstellar absorption line, Lyman-α, is caused by neutral hydrogen atoms making upward transitions from the $n=1$ to the $n=2$ level. An atom that has been excited to the $n=2$ level quickly makes a spontaneous transition back to its ground level. Since the Lyman-α photon that is re-emitted is likely to be moving in a quite different direction to the starlight originally absorbed, the overall effect is as if the photon had been scattered by the neutral atom. When we study interstellar absorption lines in the spectra of bright stars we can simply assume that this scattered light is lost into space, but in the special case of the Sun we can actually observe it. All around the sky it is possible to detect the faint glows of the 1216-Å hydrogen Lyman-α and of its 584-Å helium equivalent. This glow is called the **solar backscatter**, since it comes from sunlight that is scattered back towards the Earth by the interstellar gas. Measurements of the strengths and Doppler widths of the lines indicate that the gas has a temperature of about 10^4 K and a density of neutral atoms of 0.1 particles cm^{-3}. As we noted in Chapter 11 these conditions are what we expect in the warm neutral regions of the interstellar medium. The backscatter provides part of the evidence that the Solar System lies towards the edge of a neutral cloudlet inside the much hotter and more diffuse local bubble.

How do we distinguish between the interstellar gas and the solar wind? First, the gas that produces the backscatter is necessarily atomic, unlike the solar wind, which is ionized. Second, the interstellar gas is moving in quite a different direction to the solar wind. The Doppler shifts of the backscattered emission lines indicate that the gas is flowing into the Solar System with a speed of about 20 km s^{-1} from the direction of the constellation of Monoceros. (The flow occurs because the Solar System and the local interstellar medium are in slightly different orbits around the Galaxy. Because their densities are so low the interstellar gas and the solar wind make very little contact with each other. The neutral gas is affected by the Sun in several ways, however. First there is gravity. This force has most effect on the helium atoms, which are tugged inward as they pass the Sun, generating a 'wake' on the downwind side. Second, there is radiation pressure which acts in the opposite direction to gravity. This force is negligible for the helium atoms, because most solar ultraviolet photons have too long a wavelength to get absorbed by helium atoms, but is about the same strength as gravity for hydrogen. Third, there is photoionization which acts mainly on the hydrogen. The combined effect of radiation pressure and photoionization is to produce a zone round the Sun free of

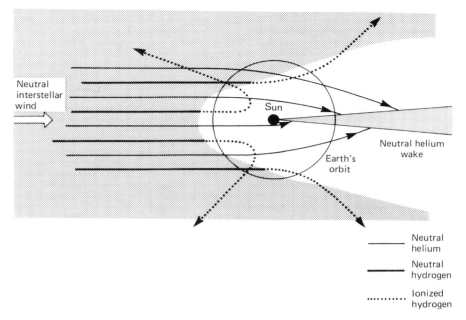

Figure 13.5. Atomic hydrogen and helium from the interstellar medium behave differently as they enter the Solar System. The neutral hydrogen atoms are ionized by the Sun and join the solar wind as it blows away from the Sun, but the helium stays neutral and is pulled into a wake by the gravitational force from the Sun.

neutral hydrogen. The size and shape of this zone fluctuates because of changes in the Sun's ultraviolet output, but its dimensions are roughly those of the Earth's orbit round the Sun.

13.5 The edge of the Solar System

For many people the edge of the Solar System is taken to be the orbit of the most distant planet. This is currently Neptune, at 30 a.u. from the Sun, but Pluto's elliptical orbit takes it out as far as 40 a.u. on occasion. (An astronomical unit [a.u.] is the mean distance of the Earth from the Sun; see Appendix B). Does this definition of the edge of the Solar System make any sense in the context of a boundary between the interplanetary and interstellar media? As might be expected, the answer is not simple.

Near the Earth the particle density of the solar wind is about 100 times that of the local interstellar medium. As the solar wind expands away from the Sun it becomes less dense, so that beyond the orbit of Saturn (10 a.u.) the particle density of the stellar wind becomes less than the local interstellar density. More important than the particle density, however, is the magnetic field, since this is inextricably tied to the motions of all charged particles. In the inner Solar System the magnetic field is generated primarily by the currents inside the Sun. In the outer regions the field is sustained mainly by the motions of the solar wind particles. The field gets weaker away from the Sun, adopting a spiral pattern that is a consequence of the Sun's rotation. Theoretical calculations indicate that somewhere well beyond the orbit of Pluto there should be a rather sharp transition from the solar-system-generated magnetic field to the interstellar magnetic field. This boundary is called the **heliopause**. It resembles a shock wave of the kind described in Chapter 4, with the important difference that the interaction between the different gas phases is via magnetic forces rather than direct collisions. Beyond the heliopause the solar wind particles merge with the general interstellar medium. Calculations put the distance to the heliopause at about 100 a.u. from the Sun – roughly three times the radius of Neptune's orbit, but less than 0.1% of the distance to the nearest star. The estimated distance to the heliopause could easily be wrong by a factor two, and is probably different in different directions from the Sun.

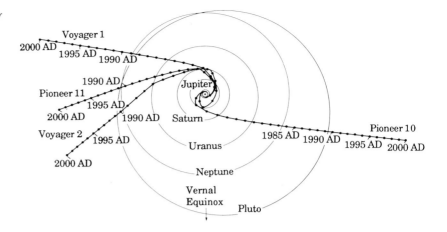

Figure 13.6. Map of the outer Solar System, showing the paths of the four spacecraft which have enough speed to escape the pull of the Sun's gravity.

The best hope of detecting the heliopause is to keep our fingers crossed. During the 1970s NASA launched four highly successful spacecraft designed to study the outer planets. The most famous of these, Voyager 2, visited Jupiter, Saturn, Uranus and Neptune. All four spacecraft have now passed the orbit of Pluto but despite their enormous distances from the Earth are still making observations of the solar wind and transmitting the data back to us by radio. Sometime during the next century each of these spacecraft will cross the heliopause. If they are still in working order, the transition should be fairly obvious. Instruments that measure the magnetic field, as well as the density, temperature and speed of charged particles around the spacecraft should register strong changes at the heliopause, though we do not know whether these changes will take place overnight, or over a period of months. The spacecraft move about 2 a.u. farther from the Sun each year; it is touch and go whether they will get to the heliopause while their radio transmitters are still in range of the Earth.

The heliopause may not be the ultimate edge of the Solar System. In 1950 the Dutch astronomer Jan Oort suggested that there are enormous numbers of unobserved comet nuclei slowly orbiting the Sun far beyond the orbit of Pluto. This collection of comets is referred to as the **Oort cloud**. The Oort cloud has never been seen, but its existence resolves a paradox concerning the ages of comets. A comet that passes close to the Sun and forms a tail is clearly being destroyed as we watch it. Halley's comet will probably survive only a few thousand years more before being evaporated away to nothing. On the other hand, comet nuclei themselves are all believed to date from the birth of the Solar System, 4.6 billion years ago. To reconcile these very different lifetimes we must suppose that Halley's comet, and others like it, were held in cold storage for billions of years before starting to make their regular visits to the inner Solar System. Oort's idea is that there is a reservoir of comet nuclei gravitationally bound to the Sun, but so far from it that they remain completely frozen solid for billions of years. Every so often a comet nucleus is nudged into a slightly different orbit by the gravitational pull of a passing star. If the new orbit takes the nucleus into the inner Solar System it will become a visible comet, and will start on the final phase of its life. Support for the existence of the Oort cloud comes from the fact that many 'new' comets are observed to have elliptical orbits that stretch up to 100 000 a.u. from the Sun.

It is estimated that the Oort cloud contains some 2×10^{12} comet nuclei. Enormous though this number seems, the total mass of all these comets is much less than the total mass of the planets. Their distances are in the range

10 000 to 100 000 a.u., reaching out nearly half way to the nearest star. The vast size of the Oort cloud has led to speculation that the Milky Way could be filled with 'interstellar comets', freed from gravitational binding to a particular star. Unfortunately, this idea may remain speculative for a long time, since, unless it passed very close to the Sun, an interstellar comet would be just a tail-less nucleus that is far too faint for any present-day telescope to detect. According to some estimates one interstellar comet could pass within a few a.u. of the Sun every 100–200 years. Such a comet would probably be recognizable by its anomalously high velocity and its unusual orbit.

14 *Geospace*

In the last chapter we described the nature of the interplanetary medium, and speculated about the way it merges with the surrounding interstellar medium. Even closer to home is a region of space in which the Earth exerts the strongest influence rather than the Sun or the Galaxy. This region of Space may be considered as the uppermost part of the Earth's atmosphere, although it bears little resemblance to the air we breathe. The region is of great technological importance, since it is here that most satellites travel, and most astronautical activities take place.

The techniques for studying the Earth's upper atmosphere have little in common with those used for studying the interstellar medium. The main tools are instruments carried in orbiting spacecraft or high-flying rockets. These instruments measure such things as the density, temperature and magnetic fields surrounding the spacecraft. The measurements are much more direct than those of the astronomer, who is forced to rely on electromagnetic waves to provide him with his information. It is fortunate that direct data are available, because the near-Earth space environment is much more complicated than the interstellar medium – at least as far as our understanding goes. Major changes can occur in less than 100 km. To understand why these changes occur we have to look at the three main factors which affect the gases in the near-Earth environment. These are the gravitational field of the Earth, the radiation from the Sun, and the magnetic field of the Earth. We describe the main effects caused by these factors in the next three sections

14.1 Gravitational equilibrium

The main force which stops the Earth's atmosphere from disappearing into space is gravity. The weight of the Earth's atmosphere at sea level is equivalent to 1 kg per square cm or 15 lbs per square inch. If we move upward through the Earth's atmosphere there is less air pressing down on us from above. This is why the atmospheric pressure and, consequently, the atmospheric density decrease with altitude (see Appendix J). For the first few kilometers in altitude the density halves every 6 km. This drop in pressure is dramatic – it means that an airliner flying at 9 km altitude (about the height of Mount Everest) is above 70% of the Earth's atmosphere. The bulk of the gas in our atmosphere is evidently confined to a very narrow layer, though, as we shall see, the tenuous outer layers of the atmosphere occupy a volume of space considerably larger than the Earth itself.

The rapidity with which the density of a gas decreases with height

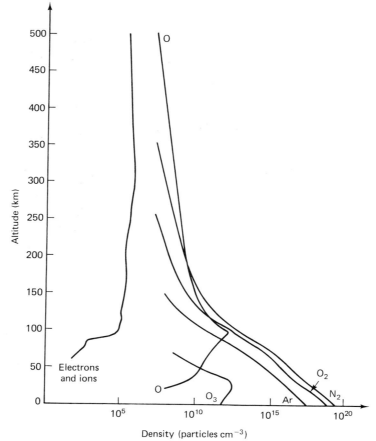

Figure 14.1. The density of the Earth's atmosphere falls with height. For the first 100 km or so the mixture of gases is similar to that at sea level. At higher altitudes gases become separated according to their molecular weights, and solar ultraviolet radiation breaks down oxygen molecules first to atoms, then to ions.

depends on the temperature; cold gases form flatter layers than hot ones. It also depends on the mean **molecular weight** of the gas. The molecular weight is the ratio of the mass of a molecule to the mass of a hydrogen atom. The air at sea level is a mixture of approximately 78% nitrogen molecules (N_2), 21% oxygen molecules (O_2) and 1% argon atoms. Their molecular weights are 28, 32 and 40, for a mean value of 29. Left to themselves gases with high molecular weight tend to 'lie low' in a flatter layer than gases of low molecular weight. This separation process is important in the upper layers of the atmosphere but at lower altitudes the density is high enough that molecular collisions mix the various different gases together very effectively. For the first 100 km or so the composition of the atmosphere remains much the same as that at sea level, but above this height some major changes start to occur. First, as we will discuss in the next section, ultraviolet sunlight is able to dissociate oxygen molecules into atoms; above about 120 km these oxygen atoms outnumber the oxygen molecules. Secondly, the density becomes so low that collisions between different types of molecules become rare. The different gases in the atmosphere therefore no longer get properly mixed. Because of their different molecular weights different gases in the atmosphere start to form layers of different thickness. The oxygen atoms, with a lower molecular weight (16), form a higher layer than the oxygen or nitrogen molecules, so that above 180 km they become the dominant constituent in Earth's atmosphere. Higher still, above around

1000 km altitude, are layers in which the lightest gases of all, helium and hydrogen, are dominant. Both these gases are so light that they slowly escape from the Earth's gravity out into space; they are replenished by new gases from the Earth's lower atmosphere. The helium is produced by the radioactive decay of uranium in the rocks of the Earth. The hydrogen atoms result from the dissociation of water molecules by ultraviolet sunlight.

Since the atmosphere is in constant motion, gases continually circulate between low and high altitudes, adjusting their pressure and density to the prevailing conditions. It is a consequence of the law of conservation of energy that gases become hotter as they are compressed and cooler as they expand. Air rising upward in the atmosphere therefore cools. The drop in temperature with height – about 7 K for every kilometer in altitude – is enough to explain why snow is more often found on mountains than in valleys. The temperature changes are also a major driving force for the enormously varied weather patterns that we enjoy or suffer from on Earth.

14.2 Dissociation and ionization

The Sun emits part of its energy as ultraviolet photons. Those with a wavelength of less than about 3000 Å have enough energy to dissociate some of the gases in the Earth's atmosphere. The most important dissociation reaction that takes place is the breaking of oxygen molecules (O_2) into oxygen atoms. As sunlight travels down through the Earth's atmosphere more and more ultraviolet photons get absorbed by oxygen molecules as they dissociate. Above an altitude of about 120 km oxygen atoms outnumber oxygen molecules, and above about 180 km oxygen atoms are the main constituent of the Earth's atmosphere. Free oxygen atoms are only generated at high altitudes; lower down in the atmosphere there are no ultraviolet photons left to break up the molecules. The number of free nitrogen atoms is insignificant at all altitudes, because only photons of energy greater than 9.7 eV can dissociate the nitrogen molecule, and very few such photons are produced by the Sun.

At intermediate altitudes of around 20–50 km the dissociation of oxygen molecules leads to the production of **ozone**, a special form of oxygen having three atoms (O_3) instead of the usual two. Ozone is formed when a newly created single oxygen *atom* collides with an oxygen *molecule*. The maximum density of ozone is at about 25 km. (Higher than this there are fewer molecules: lower than this there is less ultraviolet radiation.) Ozone molecules absorb ultraviolet photons when they themselves are dissociated, including photons which are ignored by oxygen (O_2) molecules. Since excessive ultraviolet radiation is generally harmful to living matter, we rely on the ozone layer for our own protection. Ozone is not a very stable chemical. It easily undergoes reactions with other chemicals in the upper atmosphere, and can be easily destroyed by certain chemical pollutants.

Ultraviolet photons from the Sun start to cause significant ionization of the Earth's atmosphere above about 100 km altitude. The name **ionosphere** is therefore often applied to the layers of the atmosphere between about 100 and 500 km. The most common ions in the ionosphere are those of atomic oxygen (O^+) and of molecular oxygen (O_2^+). The maximum concentration of ions (about 10^6 ions cm^{-3}) occurs at about 300 km, although even at this level the ions and their associated electrons are outnumbered 1000 to 1 by neutral oxygen atoms. Because the ionosphere is maintained by sunlight, its height, its degree of ionization, and its thickness vary with geographical position. The ionosphere also undergoes rapid changes according to the time of day, the season of the year, and the ultraviolet brightness of the Sun.

The ionosphere is important to us in two major ways. First, it protects us from solar X-rays and short-wavelength ultraviolet radiation. Second, it greatly facilitates radio communication around the world. Low frequency radio waves have difficulty in propagating through an ionized gas. They therefore become trapped in the gap between the Earth's surface and the ionosphere. In this way radio transmissions which would otherwise have been lost to space can be funneled round the curved surface of the Earth, being reflected alternately by the ionosphere and the Earth's surface. Only waves longer than some critical wavelength get reflected by the ionosphere; The critical wavelength depends on the maximum density of electrons in the atmosphere. It typically has a value of a few meters, but changes continuously with the current state of the ionosphere. For this reason, long-distance communication in the so-called 'short-wave' region of the electromagnetic spectrum is always a somewhat chancy procedure; without the ionosphere it would not be possible at all, however.

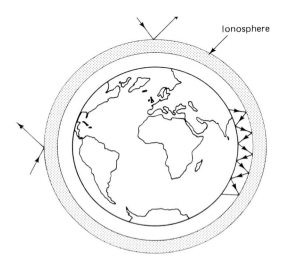

Figure 14.2. The ionosphere is opaque to wavelengths longer than some critical wavelength, usually a few meters. This presents a problem to astronomers who would like to study long-wavelength radio waves from the Galaxy, but it has the advantage of facilitating world-wide radio communication in the 'short-wave' band.

There is one group of people that is adversely affected by the ionosphere. Since the ionosphere presents a barrier to all radio waves longer than some critical wavelength astronomers are prevented from gathering data in the very lowest frequency parts of the electromagnetic spectrum. Radio astronomy from the Earth's surface has to be confined to wavelengths shorter than about 2 meters. Experiments have been carried out using low frequency antennas on a satellite orbiting above the ionosphere, but the data obtained this way have been sparse. The main problem with doing low frequency radio astronomy from space is that because of the effects of diffraction, the antenna has to be many kilometers long if it is to see any useful detail. The ideal arrangement would be to build a low frequency radio telescope on the Moon, but such a project is unlikely to be undertaken until well into the next century.

In this section we have discussed several processes which involve the absorption of ultraviolet photons from the Sun. One of the side-effects of this absorption is that solar energy is released in the upper layers of the Earth's atmosphere. The heating is significant at altitudes greater than about 15 km as a result of the absorption of ultraviolet energy by ozone molecules. The decline in atmospheric temperature with altitude that we mentioned in the previous section is reversed, and the temperature of the atmosphere increases from about 220 K (-53 °C) to a maximum of about

273 K (0°C) as we move from 15 km to 50 km in altitude. Above this height the temperature drops again as the ozone concentration decreases, but it starts to pick up again at the bottom of the ionosphere. The heating of the upper atmosphere as a result of photoionization is such that between 80 km and 300 km the temperature increases from about 200 K (-73 °C) to 1000 K or more. The maximum temperature can range between 600 and 2000 K, depending on how much ultraviolet radiation is being received from the Sun.

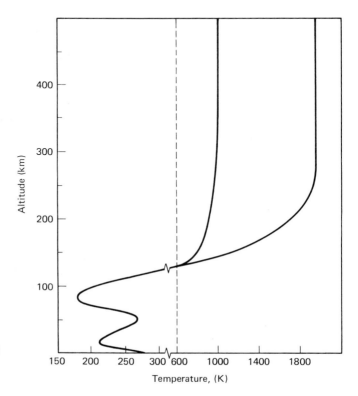

Figure 14.3. Temperature of the Earth's atmosphere. The temperature rise that occurs above 15 km is caused by the absorption of sunlight by dissociating ozone molecules. The rise above 80 km is due to photoionization of oxygen and other atoms. The maximum temperature of the upper atmosphere can vary between 600 K and 2000 K depending on the ultraviolet output from the Sun. Note that the temperature scale changes above 300 K.

The changes in temperature of the Earth's upper atmosphere lead to significant variations in its density. The changes occur because the atmosphere expands as it heats; since the only direction that the atmosphere can expand is upwards, the density at high altitude must rise. The density changes can be substantial. Between periods of sunspot minimum and sunspot maximum the mean atmospheric density at a height of 800 km can change by a factor of 100. Designers of satellite missions have to take account of the state of the Sun in ensuring that their spacecraft do not suffer so much atmospheric drag that they fall out of the sky. The drag is much greater during periods of maximum sunspot activity.

14.3 The magnetosphere

The third main influence on the Earth's upper atmosphere is its magnetic field. As we discussed in Chapter 9, a magnetic field exerts a force on any moving charged object, including the electrons and ions in the ionosphere. Effects attributable to the Earth's magnetic field start in the ionosphere and become increasingly more important with increasing altitude. Above about 7000 km – approximately the radius of the Earth – the atmosphere consists almost entirely of ionized hydrogen and the magnetic field is in total control. This region is called the **magnetosphere**.

The Earth's magnetic field is shaped something like a doughnut. From the shape of the field we can deduce that it is generated deep in the Earth's core, almost certainly by currents of molten iron. Ions and electrons are forced to travel around the magnetic field lines in helical paths. As in the interstellar medium, the magnetic field and the ionized gas effectively become attached to each other. The link works in both directions; daily and seasonal motions of the ions and electrons in the ionosphere constitute an electric current which itself generates a magnetic field. These fluctuating ionospheric magnetic fields can be detected by sensitive instruments on the surface of the Earth; they provide physicists with part of the information they need to monitor the behavior of the ionosphere.

Among the most important features of the magnetosphere are the **van Allen belts**. These are concentrations of ions and electrons which have acquired enough energy to move at speeds close to that of light. Their energies are comparable with those of interstellar cosmic rays. There are two main belts, one at about 4000 km altitude, and one at 20 000 km (about 3 Earth radii). The particles in the belts are trapped by the Earth's magnetic field, bouncing like beads on a string between the Earth's North and South poles. They are not as numerous as the photoionized ions and electrons that comprise the regular ionosphere, but their high energies make them far more conspicuous. Most of the particles originate from the Sun, having been shot out into space during the course of some kind of solar storm. Other particles are interstellar cosmic rays that have wandered into and been trapped in the magnetosphere, or particles generated by a series of nuclear explosions that were detonated in the upper atmosphere by USA and USSR military authorities in the 1960s. The van Allen belts can cause serious problems for spacecraft, since the highly energetic particles can cause havoc with sensitive instruments. Most satellites, unless they are specifically

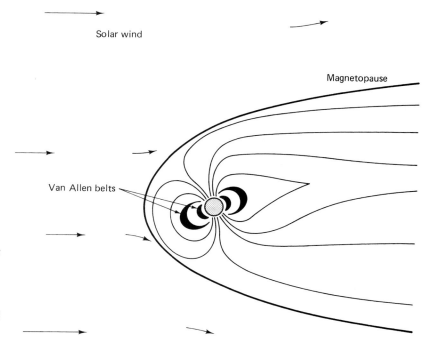

Figure 14.4. The highest layers of the Earth's atmosphere are strongly affected by the Earth's magnetic field. Charged particles moving at relativistic speeds become trapped, forming the van Allen belts. Farther out is the magnetopause – the boundary between the Earth's space, and the interplanetary medium.

designed to study the van Allen belts, avoid them if they can, staying in orbits that are either well below them or well above them.

The magnetosphere acts in many ways like the heliosphere of the Sun. It surrounds the Earth like an invisible bubble. The boundary between it and the interplanetary medium, which is called the **magnetopause**, is only a few hundred kilometers thick; at its nearest, the magnetopause is at a distance of about 10 Earth radii. Inside the magnetopause the magnetic field with its associated van Allen particles is linked to the Earth; outside the magnetopause we find the solar wind, which is linked to the much more irregular magnetic field of the Sun. Because the solar wind is streaming past the Earth at high speed, the magnetosphere gets pushed into a teardrop shape, extending much farther from the Earth on the side away from the Sun. If it were possible to view the Earth from space with eyes that could see the magnetosphere our planet would look something like a comet.

Perhaps the most important lesson we can learn from this brief look at the properties of the Earth's upper atmosphere is that low density gases, particularly if they are ionized, can behave in some very complicated ways. We should be aware that many of the ideas we have about the interstellar medium may grossly underestimate the complexity of the processes actually taking place in the Galaxy. The areas where our ideas are most liable to be an oversimplification are those in which rapid changes can occur in time. Given what we have learned about changes to the ionosphere that arise as a result of the Earth's daily rotation or the Sun's surface condition, we must be particularly wary of placing too much confidence in our understanding of how gases are expelled from stars, and how stars form out of the interstellar medium. Closer home still, the difficulty that meteorologists have in predicting the weather within the lowest 10 km of the Earth's atmosphere should remind us to be humble in our claims to understand the nature of interstellar clouds.

15 Intergalactic matter

For the finale of this book we leave the parochial bounds of the Earth, the Solar System and even the Galaxy itself. The questions we have left to the end are those concerned with the existence and nature of **intergalactic matter**. Is there gas or dust between the galaxies, and if so, what is it like, and how can we find it?

In this chapter our interest in intergalactic matter is to complete our inventory of what fills the spaces in our Universe. For many astronomers, however, the importance of intergalactic matter comes from the roles it plays in cosmology. These are fourfold. First, astronomers who study distant galaxies for clues about the history of the Universe need to worry whether their data is being distorted or limited by intervening matter. Second, there is the possibility that present-day intergalactic matter may include gas left over from the era when galaxies were formed. Such gas might be very different from present-day interstellar matter since it would presumably not have been enriched by elements that had been synthesized in stars. Third, the production or absorption of intergalactic matter over periods of billions of years could influence the way that galaxies evolve. Fourth, there is the possibility that all of space could be filled by some kind of 'dark matter' that has hitherto escaped direct detection by any kind of telescopes.

Because it has such a low density, intergalactic matter is much harder to observe than interstellar matter, and only a very few of the techniques we discussed in this book have so far been successfully applied. All they have yielded as yet are a few scattered threads of evidence. Three such clues are described in the next three sections, but it must be borne in mind that these clues refer to quite different samples of gas. We are a long way from being able to put these disparate data into a broad consistent picture of the kind we have been able to assemble for interstellar matter.

15.1 Fountains, bridges and starbursts

The usual instinct of an astronomer when faced by a new phenomenon is to concentrate first on the examples that are nearest to home. In the present context this means looking for evidence for intergalactic matter in the vicinity of the Milky Way.

Such evidence as there is for nearby extragalactic gas comes from surveys for the 21-cm emission line of atomic hydrogen. In some parts of the sky well-defined clouds of atomic gas are seen high above and below the galactic plane. These clouds have Doppler shifts much greater than would be expected for gas in the solar neighborhood; for this reason they are often

170 INTERGALACTIC MATTER

referred to as **high-velocity clouds**. It is very difficult to ascertain the distance to these clouds; as a result their location and nature is quite controversial. Different high-velocity clouds may have different origins. Among the suggestions for their nature are:

(1) **Intergalactic gas** entering the Milky Way for the first time. A problem with this idea is that at least some of the clouds have relatively normal chemical abundances, with many more heavy elements than would be expected from gas that had never previously been part of a galaxy.

(2) Gas previously ejected from the plane of the Galaxy and now falling back towards it. This idea is referred to as the **galactic fountain** and was mentioned in Chapter 12. The gas could have been thrown up from supernovae in regions of OB star formation, or from some kind of explosion in the center of the Galaxy. A problem with this idea is that some of the clouds are falling faster than would be expected based on the gravitational field of the Galaxy's disk.

(3) Gas pulled out of the Milky Way by the gravitational pull of another galaxy. There is little doubt that some of the high-velocity gas near the Milky Way Galaxy is shaped by the gravitational attraction of the

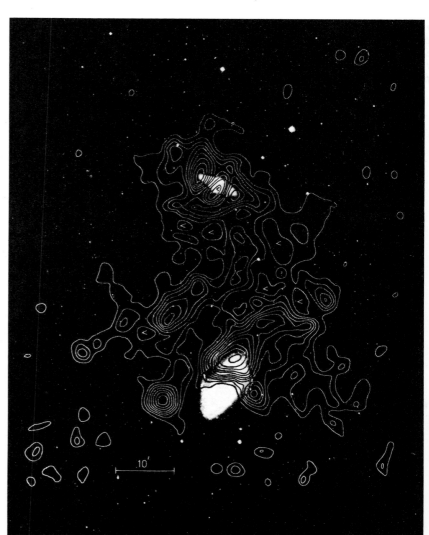

Figure 15.1. Map of the atomic hydrogen clouds in the vicinity of the M81 group of galaxies. A near-collision between two bright galaxies about 2×10^8 years ago has caused a bridge of atomic hydrogen gas to be pulled into space from out of the disks of the galaxies.

Magellanic Clouds – the two small galaxies that are the Milky Way's closest companions. Gas clouds that are pulled out of one galaxy by another are sometimes referred to as **tidal bridges or tails** by analogy with the Moon's pull on the oceans of the Earth. Bridges refer to gas clouds that join galaxies; tails refer to gas streams that point in other directions. Tidal bridges are quite common in other clusters of galaxies, and, indeed, are easier to recognize at a distance than when they almost surround us, as does the bridge which connects our Galaxy to the Magellanic Clouds.

Interestingly, near-collisions between galaxies can affect their nuclear regions as well as their outer layers. If galaxies get too close together interstellar gas in the spiral arms of one galaxy may be pulled out of its orbit by the gravitational force from the other galaxy. If the gravitational pull is in the right direction the gas's speed of revolution may be slowed enough for some of the gas to fall to the center. The rapid build-up of gas near the nucleus of the galaxy results in the rapid formation of millions of new stars in a very short time. This process is called a **starburst**. The combined power emitted by all these new stars can be the equivalent of tens of billions of Suns, but because there are substantial amounts of dust in the starburst region most of this power comes out at infrared wavelengths.

15.2 Hot gas in galaxy clusters

The most unambiguous evidence for intergalactic matter has come from the discovery of X-ray emission from clusters of galaxies. Galaxy clusters are the largest concentrations of matter which can be easily distinguished

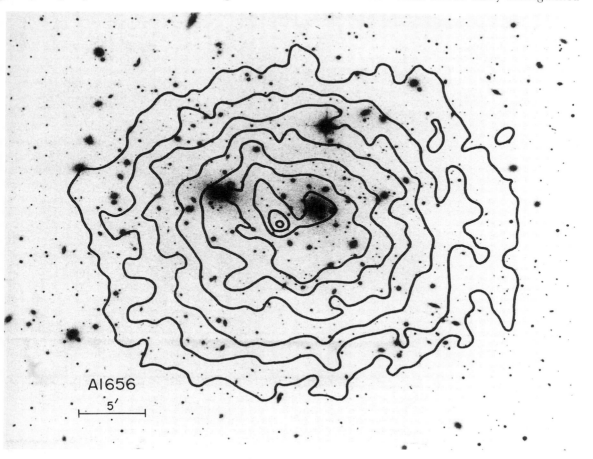

Figure 15.2. The Coma cluster of galaxies as seen at visible wavelengths and in X-ray emission. The X-rays are produced by a very hot thin intergalactic gas that is trapped by the gravitational field of the galaxy cluster.

on photographs of the sky. Rich clusters of galaxies, such as the Virgo cluster, contain several hundred galaxies. Poorer clusters, such as the 'Local Group' in which we reside, contain less than 50. Some galaxy clusters have regular, well-defined shapes with sizes a few million parsecs across, while others have a more irregular appearance and may merge into their neighbors.

X-rays have now been detected from nearly 100 clusters. Maps of these clusters made by the Einstein Observatory satellite show that the X-rays come from the whole cluster, not from one or two individual galaxies. Another very important result is derived from X-ray spectroscopy of the clusters. They show a strong emission line of highly ionized iron at a wavelength corresponding to an energy of 7 keV. This 7-keV iron line (actually a blend of several lines of Fe^{23+} and Fe^{24+}) provides us with four crucial pieces of information. First, it shows that iron atoms exist in the intergalactic medium of these clusters. The iron abundance is roughly half that of the Sun, indicating that at least some of the gas almost certainly has been processed through stars sometime in its past. Second, it shows that the emission process for the X-rays must be thermal free–free emission from hot gas, rather than synchrotron emission or some other non-thermal process. Third, by comparing the strength of the iron line with some other weaker lines, we can determine that the temperature of the gas is around 10^8 K, and its density is around 10^{-3} atoms cm^{-3}. At this temperature the gas has to be completely ionized; it is somewhat like the coronal gas in the halo of the Milky Way, but hotter. Fourth, once we know the density and size of this hot gas we can calculate its mass. Remarkably, we find that rich clusters can have more than 10^{14} $M_\odot$ of hot gas; the mass of intergalactic matter within a cluster can equal that of all the stars in all the galaxies in the cluster.

There is independent evidence that galaxy clusters sometimes contain substantial amounts of gas. The evidence comes from a study of radio galaxies. A radio galaxy is a rare type of galaxy which emits thousands of times more radio power than normal. The primary source of energy in a radio galaxy is thought to be a black hole at its nucleus. By some mechanism which we do not fully understand the nucleus is able to eject vast quantities of high-energy plasma into extragalactic space. These streams of plasma are emitted as narrow jets in opposite directions. In isolated radio galaxies these jets move in straight lines for tens or hundreds of kiloparsecs until their energy is converted to radio waves by the synchrotron process. They therefore have a linear appearance with bright patches of radio emission on either side of the visible galaxy. If the radio galaxy is part of a cluster the situation is more complicated. As the galaxy moves in its orbit through the cluster the plasma streams collide with the intergalactic gas and are blown behind it, like smoke from a moving train (see Figure 15.3). Radio galaxies inside clusters therefore have a much more complicated structure than isolated radio galaxies. Analyses of the patterns produced by radio galaxies as they move through clusters are consistent with particle densities of around 10^{-3} cm^{-3}, as estimated from the X-ray emission.

Intergalactic gas in clusters becomes easier to understand if we examine some of the consequences of its high temperature. First, we note that the gas is hot enough to support itself against the gravitational pull of the cluster itself. Another way of saying this is that the half-height of the cluster's 'atmosphere' is about equal to its radius. Second, the gas radiates very inefficiently at these temperatures, for reasons we discussed in Chapter 11. Except at the very center of the cluster, the gas cools so slowly that its 10^8

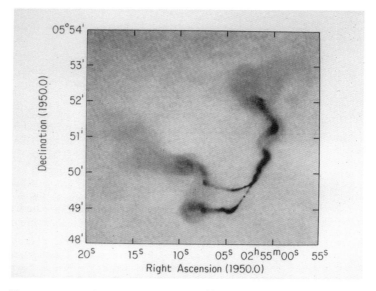

Figure 15.3. The radio source 3C 75 lies in a rich cluster of galaxies. The plumes of radio-emitting plasma are blown into strange-shaped trails by the intergalactic gas in the cluster. This particular object is unusual in having two galaxy nuclei that emit radio jets in the same galaxy cluster.

K temperature hardly changes in 10^{10} years – the age of the Universe. These two results imply that the intergalactic medium in clusters is a very stable, long term phenomenon, and that it could have accumulated there at almost any time in the past.

The origin of the cluster gas is a mystery. The earliest explanation was that it is gas that has been attracted from outside the cluster by the combined gravitational forces of the galaxies. The high temperature results from the acceleration the atoms receive as they fall into the cluster; in other words the thermal energy is derived from the gravitational potential energy of the infalling gas. The problem with this explanation is the presence of iron; gas left over from the Big Bang should contain nothing but hydrogen and helium. If the gas does not come from outside the cluster it must come from the galaxies within it. Perhaps the gas comes from processes like that in the Milky Way's galactic fountain, with matter being blasted out of the galaxies by supernova explosions. Alternatively the gas might be stripped out of the galaxies during near-collisions. The problem with either of these explanations is the enormous quantity of gas needed; we must postulate that during the lifetime of a galaxy half its mass must be ejected into intergalactic space. We do not see mass loss on this scale going on in our Galaxy now, but we cannot exclude the possibility that conditions were much more violent in the past.

A final point of interest about intergalactic gas in clusters is that although most of the gas stays hot indefinitely, cooling of the gas can and does occur at the center of a cluster, where the densities are highest. As this gas in the center cools its density increases, causing it to sink and fall towards the galaxy closest to the center of the cluster. This process is referred to as a **cooling flow**. In some clusters the cooling flow can be seen as filaments extending from the central galaxy. These filaments emit hydrogen recombination lines. Because it gets to collect this infalling gas, the central galaxy in a cluster often stands out as having a much higher luminosity than its companions. The rate of infall may be as much as 100 $M_\odot$ yr^{-1}; the total amount of material accreted by the central galaxy in its lifetime can therefore be around 10^{12} $M_\odot$ – more matter than most galaxies contain altogether. The fate of the infalling gas as it accumulates in the central galaxy is something of a mystery. If it stayed gaseous we would be able to detect it as H^0, or H^+ or H_2, which we do not. A more likely fate is that it gets turned into stars, but the direct evidence for this is slim. If 100 $M_\odot$ yr^{-1} were turned into new stars under the conditions that apply in the spiral arms of our Galaxy we would expect so many luminous OB stars to be formed that the galaxy should be full of H^+ regions and much more luminous than it actually appears. Perhaps in the special conditions of a cooling flow something prevents high mass stars from forming.

15.3 Quasar absorption lines

How can one search for intergalactic gas outside galaxy clusters? A good place to start would be to look for the Lyman-α absorption line in the spectra of distant galaxies, since this is the strongest absorption feature in the interstellar medium. At first glance the search looks hopeless. For one thing, the Lyman-α line at 1216 Å can be observed only by using satellites. For another, absorption and emission of Lyman-α radiation occurs widely *within galaxies*, so that the discovery of an absorption line at 1216 Å in the spectrum of a galaxy does not necessarily tell us anything useful about intergalactic matter. The way to get around both of these problems is to observe distant quasars. Because of the expansion of the Universe, the more distant a quasar is, the faster it is moving away from us, and the greater is the Doppler redshift of the lines in its spectrum. The quasars which are useful are those which are moving so fast that the 1216-Å Lyman-α line is redshifted into the visible waveband, between 4000 and 7000 Å. Such a quasar would be moving at around 90% of the speed of light.

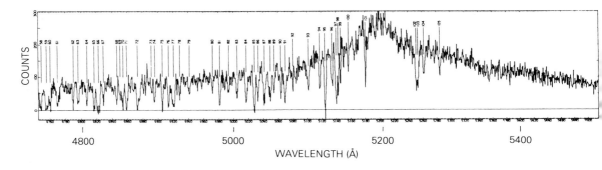

Figure 15.4. Spectrum of the quasar PKS 2126–158. Almost all the absorption lines are Lyman-α lines of hydrogen – the transition from $n=1$ to $n=2$. Each transition has a different redshift, indicating that the clouds that are doing the absorbing are scattered at all distances along the line of sight to the quasar.

A spectrum of a high-redshift quasar is shown in Figure 15.4. The very broad Lyman-α emission feature is produced by the quasar itself and tells us nothing about the intergalactic medium. The dozens of narrower absorption lines at wavelengths shortward of this broad feature are certainly not produced in the quasar, however, since their wavelengths do not match those of any known element moving near to the quasar's velocity. By far the most likely explanation for these absorption lines are that they are all Lyman-α lines occurring at different redshifts. As far as we can tell, absorptions can occur at any redshift between zero and the wavelength of the quasar itself. The implication of this result is that there are clouds containing atomic hydrogen spread out throughout space between here and the distant quasars.

The existence of atomic gas in intergalactic space is at first sight surprising. Bright ultraviolet-emitting quasars fill intergalactic space with so much radiation at $\lambda < 912$ Å that all intergalactic gas ought to be photoionized. The only way out of this dilemma is to assume that the clouds seen by their Lyman-α absorption lines are, in fact, photoionized H^+ regions that behave something like those discussed in Chapter 5. The neutral atoms which give rise to the Lyman-α absorption lines are a tiny minority compared to the ionized atoms (about one part in 10^4) and exist only temporarily in the intervals between a recombination of an ion and an electron and its re-ionization by another ultraviolet photon. The diameters of the clouds are very uncertain, but each is perhaps about the size of a galaxy. The density of the gas in the clouds is estimated to be around 10^{-4} or 10^{-3} ions cm^{-3}. Gas temperature is around 30 000 K. The absence of absorption lines from elements other than hydrogen indicates that this type of intergalactic gas does *not* appear to have been processed through stars; helium is probably present, but is hard to observe for technical reasons.

Knowing the size, density and temperature of these clouds we can examine their state of equilibrium using the same ideas as we used when considering the formation of a star from an interstellar cloud. We find that the internal gravitational forces trying to hold these clouds together are not as strong as the thermal pressure trying to make them expand. If left to themselves, therefore, these intergalactic clouds would simply expand outward and lose their identity. To hold them firm against dissipation we need an outside pressure holding them in. One possibility for such a pressure is a **diffuse intergalactic medium** with a density of around 10^{-5} atoms cm^{-3}, and a temperature of around 3×10^5 K. Whether such a diffuse intergalactic medium fills all of space, or only certain parts of it we do not know.

Can we get evidence for intergalactic matter in any other way? Searches for dust are not promising. Gas that has not been processed through stars is unlikely to have any dust in it, since dust is made mainly out of heavy elements. Even if it exists it will be so cold that it will give off very little radiation, and so diffuse that it will produce very little reddening. Another approach is to look for X-rays, like those found from galaxy clusters. A background of X-ray emission has been found from the whole sky, but its spectrum is wrong, and most astronomers believe that the X-rays come from distant unusual galaxies rather than from the intergalactic medium. All in all, the evidence in favor of the diffuse intergalactic medium is not strong, and an alternative explanation for the stability of intergalactic clouds will be described in the next section.

15.4 Dark matter

For most of this book we have discussed ways of detecting and studying matter by means of photons. There is a quite different method that can be used to detect matter, namely the gravitational force it exerts. Several important discoveries have been made this way. One was the planet Neptune whose position was predicted on the basis of the gravitational pull it was exerting on Uranus. Another was the faint white dwarf star which orbits Sirius; its existence was deduced from small periodic wobbles in Sirius's position in the sky. In both these cases motions of an object of known mass were used to deduce the strength of the gravitational force from an unknown source.

The same principle can be applied statistically to groups of stars or galaxies. If a star cluster is neither contracting nor expanding there has to be a balance between the gravitational forces pulling the cluster together and the random motions of the stars tending to make it expand. Specifically, a law called the **virial theorem** states that the average gravitational energy of the stars in the cluster has to equal twice their average kinetic energy. Mathematically the situation is very similar to the balance between gravity and gas pressure in a star or in a cloud, except that the moving atoms are replaced by moving stars. We can make a direct test of the virial theorem if we know the size of a cluster, the total mass of the cluster and the average speed of its members. We can get the size by measuring the cluster's angular size and distance, the mass by counting the number of objects in it and assuming an average stellar or galactic mass, and the average speed by measuring a number of Doppler shifts.

When we apply the virial theorem to different astronomical situations we get a variety of results. For old star clusters (often referred to as globular clusters) in our Galaxy we get good agreement between theory and observation, supporting the idea that these clusters are stable – neither expanding nor contracting. For young star clusters in the Galaxy we find that the kinetic energy exceeds half the gravitational energy. This is to be expected for new stars expanding away from the sites of their birth. The real surprise comes when we examine clusters of galaxies. We find that in almost every case the kinetic energy of the galaxies in the cluster is far greater than the gravitational energy. The discrepancy is usually a factor of about 10.

How are we to interpret this disagreement? It is most unlikely that the galaxy clusters are all actually expanding at the rate implied by their Doppler shifts. If this were the case galaxies from different clusters would have all got mixed up with each other billions of years ago. The clusters themselves would no longer have any identity and the Universe would be smoothly filled with galaxies.

A more likely explanation for the breakdown of the virial theorem is that we have somehow underestimated the gravitational energy in the galaxy clusters. This means we have failed to account for all the mass in it. The discrepancy is large; to satisfy the virial theorem we must postulate that galaxy clusters contain about 10 times more mass than we can easily see. For a while astronomers referred to this discrepancy as 'missing mass'. As they became convinced that something had to exist in these clusters to satisfy the laws of physics they came to prefer the term 'dark matter'. The nature of dark matter is one of science's greatest current mysteries.

Some of the dark matter may be associated with the galaxies themselves. Stars in the disks of some spiral galaxies are moving faster in their orbits than can be explained by the gravitational forces from the stars we see. Many astronomers conclude that spiral galaxies contain large quantities of dark matter in a roughly spherical halo that is at least as large as the visible disk. If spiral galaxies in general have halos that have about 10 times the mass of their stars we could explain both the anomalies in their rotation, and the stability of galaxy clusters.

A quite different argument in favor of dark matter comes from cosmology. For the last 10–20 billion years following the Big Bang, the Universe has been expanding. As the galaxies move apart they are subject to a deceleration due to their mutual gravitational attraction. The amount of the deceleration depends on the average density of matter (visible plus dark) in the Universe. If the density is above the **critical density**, (corresponding to about 10^{-4} hydrogen atoms cm^{-3}), the Universe will eventually stop expanding and will collapse in on itself: if the average density is below this value the Universe will expand forever. The measured average density of visible matter in the Universe, which is obtained by estimating the average number of stars in a galaxy and the average separation between galaxies, is about 1% of the critical density. If visible matter were the only matter the Universe would expand forever.

There is no overriding reason that we know of why the Universe should not expand forever. Nevertheless the theoretical understanding of the circumstances surrounding the Big Bang and the subsequent formation of the galaxies are now quite sophisticated. For reasons that are beyond the scope of this book several of these theories become much simpler if it can be assumed that the average density of the Universe is exactly equal to the critical density. Since one of the fundamental beliefs of all scientists is that the Universe is governed by simple principles, the idea that there is *100 times* more dark matter in the Universe than visible matter is an appealing one. Cosmologists wrestling with the problems of forming galaxies in the early Universe find their job easier if dark matter exists. It can also help explain the otherwise puzzling large scale uniformity of the Universe.

What is dark matter made of? To answer this question we must work by elimination. One possibility is that dark matter is made up of very faint low mass stars called **brown dwarfs**. The trouble here is that if there are enough brown dwarfs in the Galaxy to provide for all the 'dark matter' we are looking for we should see substantial numbers in our own neighborhood. We have not done so. If the dark matter were intergalactic *hydrogen and helium gas* it would be detectable by one or other of the techniques we have discussed in this book. For example, H^0 would show the 21-cm line, hot H^+ would show X-ray emission, and cooler H^+ would reveal itself by the sorts of quasar absorption lines we discussed in the previous section. H_2 is most unlikely to exist because it would be dissociated to H^0. *Solid particles* have also been considered and rejected. Although in the Earth's environment hydrogen solidifies at 14 K, in the low pressure of intergalactic space it quickly evaporates back to a gas. Solids made of heavy elements, such as rocks or planets, would certainly be invisible to us in intergalactic space, but would create havoc with everything we think we know about nucleosynthesis and the relative abundances of the elements. There does not seem to be any way of filling space with 'ordinary' atoms or ions to the tune of 10 times the mass of visible matter. Although they are extremely dark and can be extremely massive, *black holes* also present problems. For one thing they are hard to make. The only well-established way of making a new black hole, as opposed to enlarging an old one, is during a supernova explosion. Since less than 50% of a star becomes a black hole during a supernova explosion we would have to push interstellar gas through the star formation/supernova cycle many times to get 99% of the Universe's mass as black holes. So much nucleosynthesis would produce far more helium and heavy elements than we actually observe.

Measurements of a different sort – namely the present day abundance of deuterium – allow us to place a different kind of restriction on the nature of dark matter. By performing computer simulations of the nuclear reactions that took place in the first few minutes after the Big Bang astronomers can place limits on the present-day density of what is called **baryonic matter.** Baryonic matter is matter that is made up of **baryons** – a class of elementary particle that includes protons and neutrons, but excludes electrons and photons. Baryons make up most of the mass of an atom, so in terms of mass, baryons dominate the visible matter in the Universe. As we discussed in Chapter 10, deuterium – the heavy isotope of hydrogen – was produced and then largely destroyed during the first few minutes after the Big Bang. Detailed calculations show that the amount of deuterium left at the present time depends critically on the average density of baryons in the Universe during those first few minutes. By using the measured present-day deuterium abundance and calculating the amount of expansion that has taken place since nucleosynthesis ended, we find that the present-day density of baryonic matter is about 10% of the critical density. The implication of this result is either that the Universe is below its critical density and will expand for ever, or that some kind of mysterious non-baryonic matter is responsible for most of the mass of the Universe.

To find this new matter astronomers have had to turn to advice from theoretical physicists who are concerned with the nature of matter on the smallest scale. Their highly mathematical theories often require the invention of as-yet undiscovered particles. Some of these, like the **axion**, the **gravitino** and the **photino** have a mass less than that of the electron. Others such as the **pyrgon**, the **maximon**, and the **preon** are far more massive than the proton. Members of the latter group are sometimes referred to as '**wimps**', for **w**eakly **i**nteracting **m**assive **p**articles. None of the particles suggested so far is entirely satisfactory; the low mass particles are useful for explaining the largest structures in the Universe, but the high mass particles do a better job on the smaller scale sizes.

Perhaps we will one day find the ideal particle that can clarify all the mysterious unexplained gravitational effects in the Universe. On the other hand, let us remember that it took a dozen chapters of this book to describe the different kinds of interstellar matter in our Galaxy. It is quite possible that intergalactic space is filled with a range of 'dark' phenomena that is every bit as broad and as rich as the range of 'bright' phenomena we have discovered up till now.

Appendix A: Large and small numbers

Astronomers have to deal with large numbers all the time. To avoid being faced with rows of unnecessary digits they usually employ **scientific notation** for dealing with numbers greater than about a thousand. In scientific notation a large number is represented as a small number (usually between 1 and 10) multiplied by 10 raised to some power. When this is done large numbers become as easy to read as small ones.
Examples are:

100	10^2
1 000	10^3
100 000	10^5
1 000 000 000	10^9
3 200	3.2×10^3
72 400 000	7.24×10^7

We can use a similar notation when dealing with small numbers, with the difference that the powers of ten are negative. Small numbers become important when we discuss matters concerned with individual atoms or with the wavelength of light.
Examples are:

0.1	10^{-1}
0.001	10^{-3}
0.000 000 1	10^{-7}
0.056 7	5.67×10^{-2}
0.008 13	8.13×10^{-3}

Sometimes numbers look better when written as words. The two words which must not be confused are:

1 million	1 000 000	10^6
1 billion (US)	1 000 000 000	10^9

The word trillion (10^{12}) is rarely used by scientists.

Strictly speaking the word billion and trillion are ambiguous. In the United Kingdom they are supposed to refer to the numbers 10^{12} and 10^{18} respectively. However, since the British-born author of this book has never in his career ever seen an example of the UK billion in use, this book follows the conventional practice of using the more usefully-sized US billion.

Appendix B: The metric system and related units

Distance

All professional scientists use the **metric system**. The basic units of distance in the metric system are:

millimeter	1 mm	10^{-3} m	approx 1/25 inch
centimeter	1 cm	10^{-2} m	approx 2/5 inch
meter	1 m		approx 39 inches
kilometer	1 km	10^{3} m	approx 5/8 mile

Special units are needed for short distances. The units that are commonly used in astronomy are:

micrometer	1 μm	10^{-6} m	also called a micron
nanometer	1 nm	10^{-9} m	
ångström	1 Å	10^{-10} m	

In the strictest form of the metric system, the **Système International** (SI) the use of the centimeter and the ångström are discouraged in favor of the millimeter and the nanometer. Their use is so widespread among astronomers, however, that something of the flavor of the subject might be lost by avoiding their use in this book.

Astronomers use special units for objects much bigger than the Earth. Within the Solar System distances are based on the astronomical unit which is the mean distance of the Earth from the Sun.

1 astronomical unit (a.u.) 1.5×10^{8} km (93 million miles)

For larger distances the parsec is used:

parsec	1 pc	3×10^{13} km
kiloparsec	10^{3} pc	3×10^{16} km
megaparsec	10^{6} pc	3×10^{19} km

The parsec as a unit is a historical relic, and its continued popularity is something of an embarrassment to astronomers. An alternative unit for large distances is the light year – the distance that light travels in a vacuum in one year.

Light year   1 l.y. 9×10^{12} km or 0.3 pc

Despite its popularity in space travel fiction, the light year is rarely used by professional astronomers.

One reason for the popularity of the parsec is that one parsec is about the distance from the Sun to the next nearest star, while a megaparsec is roughly the distance from the Milky Way to the next nearest spiral galaxy. Neither the a.u., the parsec nor the light year are officially part of the Système International, but astronomers have never been particularly distressed by this fact.

Volume

Small volumes are measured in **cubic centimeters** (cm^3). The cubic centimeter is the same as a **milliliter** (10^{-3} liters) and is about the size of a small sugar cube.

Large volumes, such as sections of our Galaxy, are measured in cubic parsecs.

1 cubic parsec $\approx 3 \times 10^{55}$ cm^3

Mass

Masses in astronomy are usually expressed in **grams** (g)

| gram | 1 g | approx 1/28 oz |

The Système International favors the **kilogram**, often abbreviated to kilo

| kilogram | 1 000 g | approx 2.2 lb |
| metric tonne | 1 000 kg | |

For the purposes of this book all kinds of ton (metric, imperial and US) are near enough the same.

Large masses, such as those of stars and galaxies, are expressed as multiples of the mass of the Sun – usually written as $M_\odot$ and referred to as the **solar mass**.

| solar mass | 1 $M_\odot$ | 2×10^{33} g |

Time

Like everyone else astronomers use **seconds** (s) to measure short times and **years** (yr) to measure long times. A useful conversion is:

1 year $\approx 3 \times 10^7$ s

The age of the Universe is about 16×10^9 years.

Frequency

Frequency is measured in hertz (Hz)

1 hertz	Hz	= 1 cycle per second
1 kilohertz	kHz	= 10^3 Hz
1 megahertz	MHz	= 10^6 Hz

Energy

The basic units of energy in the metric system are the **erg** and the **joule** (10^7 ergs), but when we consider quanta of energy emitted or absorbed by individual atoms we usually use a smaller unit called the **electron volt** (eV).

1 eV = 1.6×10^{-12} erg

For somewhat larger energies we use the **kilo-electron volt (keV)**, the **mega-electron volt (MeV)** and the **giga-electron volt (GeV)**.

1 keV = 10^3 eV
1 MeV = 10^6 eV
1 GeV = 10^9 eV

Power

The most common metric measure of power is the **watt** which is one joule per second. In astronomy we often want to compare the power object of a star with that of the Sun, so we measure power output in terms of **solar luminosity** (usually written $L_\odot$).

1 $L_\odot$ = 4×10^{26} watts

Unless otherwise specified the term 'power output' includes radiation at all wavelengths.

Temperature

Temperature measurements in astronomy employ the **Kelvin** or absolute scale. The logic behind this system is outlined in Chapter 3. The symbol for a degree kelvin is simply K. In the Kelvin temperature scale ice melts at 273 K, and water boils at 373 K; as in the **Celsius** or **centigrade** scale there are 100 degrees between melting and boiling. The **Fahrenheit** scale, in which ice melts at 32° and water boils at 212° is never used in astronomical research.

Angles

There are 360 **degrees** (°) to a circle. Small angles are measured in **arcminutes** (') or **arcseconds** (").

1° = 60'
1' = 60"

One arcsecond is about the angular diameter of a small coin at a distance of a few kilometers.

The diameter of the full moon is about $\frac{1}{2}$° (30'). The smallest detail that can be seen by the unaided human eye is about 1'.

Some applications make use of the **radian** (rad)

1 rad = $(180/\pi)° \approx 57.3°$

Appendix C: Greek letters

Name	Letter	
Alpha	A	α
Beta	B	β
Gamma	Γ	γ
Delta	Δ	δ
Epsilon	E	ε
Zeta	Z	ζ
Eta	H	η
Theta	Θ	θ
Iota	I	ι
Kappa	K	κ
Lambda	Λ	λ
Mu	M	μ
Nu	N	ν
Xi	Ξ	ξ
Omicron	O	o
Pi	Π	π
Rho	P	ρ
Sigma	Σ	σ
Tau	T	τ
Upsilon	Y	υ
Phi	Φ	φ
Chi	X	χ
Psi	Ψ	ψ
Omega	Ω	ω

Appendix D: Wavelength, frequency and energy

The frequency (ν) and wavelength (λ) of an electromagnetic wave are related by the equation

$$\lambda = c/\nu \tag{D.1}$$

where c is the velocity of light.

The energy of a photon (E) is given by

$$E = h\nu \tag{D.2}$$

where h is Planck's constant. If the first two equations are combined we obtain:

$$E = hc/\lambda \tag{D.3}$$

The energy of a photon is thus inversely proportional to its wavelength.

When an atom makes a transition from an upper energy level E_2 to a lower energy level E_1 the wavelength of the emitted photon is given by

$$\lambda = hc/(E_2 - E_1) \tag{D.4}$$

Appendix E: Selected chemical elements

Name	Symbol	Atomic number	Atomic weight	Ionization potential (eV)
Hydrogen	H	1	1, 2	13.6
Helium	He	2	4, 3	24.6
Lithium	Li	3	7	5.4
Beryllium	Be	4	9	9.3
Boron	B	5	11	8.3
Carbon	C	6	12, 13	11.3
Nitrogen	N	7	14	14.5
Oxygen	O	8	16, 18	13.6
Fluorine	F	9	19	17.4
Neon	Ne	10	20	21.6
Sodium	Na	11	23	5.1
Magnesium	Mg	12	24	7.6
Aluminum	Al	13	27	6.0
Silicon	Si	14	28	8.1
Phosphorus	P	15	31	10.5
Sulfur	S	16	32	10.4
Chlorine	Cl	17	35	13.0
Argon	Ar	18	40	15.8
Potassium	K	19	39	4.3
Calcium	Ca	20	40	6.1
Scandium	Sc	21	45	6.5
Titanium	Ti	22	48	6.8
Vanadium	V	23	51	6.7
Chromium	Cr	24	52	6.8
Manganese	Mn	25	55	7.4
Iron	Fe	26	56	7.9
Cobalt	Co	27	59	7.9
Nickel	Ni	28	58	7.6
Copper	Cu	29	63	7.7
Zinc	Zn	30	64	9.4
Silver	Ag	47	107	7.6
Iodine	I	53	127	10.5
Gold	Au	79	197	9.2
Mercury	Hg	80	202	10.4
Lead	Pb	82	208	7.4
Uranium	U	92	238	6.0

Column 4 gives the atomic weight of the most common isotope. A second number in column 4 indicates a rare isotope that is of specific astrophysical interest.

Appendix F: The Doppler effect

The change in wavelength ($\triangle\lambda$) of a spectral line with wavelength λ is given by

$$\triangle\lambda/\lambda = v/c \qquad (F.1)$$

where v is the radial velocity of the object emitting the radiation. The Doppler shift can also be expressed in terms of the frequency (ν)

$$\triangle\nu/\nu = v/c \qquad (F.2)$$

A motion *away* from the Earth produces an *increase* in λ and a *decrease* in ν. Both these formulae become inaccurate for velocities (v) close to the speed of light. The typical velocities of gas clouds in our Galaxy are a few tens of kilometers per second, so the fractional changes in wavelength are only of the order of 0.01%.

Appendix G: Temperature, energy and pressure

For a low density gas composed of single atoms the relationship between the thermal energy of a gas (U) and temperature (T) is

$$U = 3NkT/2 \tag{G.1}$$

where N is the number of atoms it contains and k is a fundamental number called **Boltzmann's constant**. The energy of the gas atoms is in the form of kinetic energy of motion, and so can be expressed in terms of the 'root-mean-square' average velocity (v_{rms}) of the gas atoms and their individual masses (m).

$$U = \tfrac{1}{2} N m v_{rms}^2 \tag{G.2}$$

From these two equations the relationship between temperature and rms velocity becomes:

$$v_{rms} = \sqrt{(3kT/m)} \tag{G.3}$$

The velocity v_{rms} is an average for the atoms' motions in all three dimensions. If the motions are random, one third of an atom's kinetic energy will be taken up in each dimension. The rms *radial velocity*, which is the only part of the total velocity that can be measured by the Doppler shift, is then given by:

$$v_r = \sqrt{(kT/m)} \tag{G.4}$$

The pressure (P) of a volume (V) of gas may be derived from the ideal gas law:

$$PV = NkT \tag{G.5}$$

By combining equations (G.1) and (G.5) we see that there is a close connection between the pressure and the energy density (U/V) of a gas

$$P = 2U/3V \tag{G.6}$$

Some of these relationships become more complicated for molecules and for ionized gas, but the same principles apply.

Appendix H: Thermal radiation

The thermal radiation from a black body at a temperature T is given by the equation:

$$Bv(T) = \frac{2hv^3}{c^2[\exp(hv/kT) - 1]} \tag{H.1}$$

This expression is known as **Planck's law**. Bv is the **brightness** of the object, which is measured in units of ergs cm^{-2} s^{-1} Hz^{-1} rad^{-2}. The curves drawn in Figure 3.8 are graphs of this equation.

Planck's law can be used to derive two other important equations:

Stefan's law states that the total power (W) radiated by a black body with surface area A at all wavelengths is given by

$$W = \sigma A T^4 \tag{H.2}$$

where σ is **Stefan's constant**.

Wien's law states that the wavelength of the maximum brightness of a black body is inversely proportional to its temperature:

$$\lambda_{\max} T = \text{constant} \tag{H.3}$$

A simple physical understanding of Wien's law can be gained if we make the simplifying assumption that most of the photons a body radiates have energies roughly equal to the average energy of an atom in the body. If we equate the average energy of an atom (U/N) to the energy of a photon (E) we obtain from equations (D.3) and (G.1) that

$$3kT/2 \approx hc/\lambda \tag{H.4}$$

This equation can be turned round to give

$$\lambda T \approx 2hc/3k \tag{H.5}$$

which is an approximate form of Wien's law.

Appendix I: Galactic rotation

Consider a star of mass m in a circular orbit around the edge of a galaxy of radius R and mass M. For simplicity we assume that the gravitational force from all the stars in the galaxy is equal to that of the same mass M at the center of the galaxy. This assumption is perfectly valid if the galaxy has a spherical shape, and is a reasonable approximation for a disk. Newton's inverse square law of gravitation tells us that the force F toward the center of the galaxy is

$$F = GMm/R^2 \qquad (I.1)$$

where G is the gravitational constant

The centrifugal force that maintains the star in its orbit is

$$F = mv^2/R \qquad (I.2)$$

where v is the star's velocity.

Combining these equations gives

$$M = v^2 R/G \qquad (I.3)$$

If we can measure the orbital velocity G of a star (or gas cloud) at a radius R from the galaxy's center we can calculate the mass M of the galaxy. Strictly speaking what this calculation gives us is the mass *within* a radius R of the center. By measuring v at different values of R we can determine how the mass density of the galaxy varies with radius.

Appendix J: Atmospheres and gravity

Consider a vertical column of gas with cross-sectional area A that is subject to a downward gravitational acceleration g. The gas has a mass density ρ which varies with the height h.

The mass δm of the gas between h and $h + \delta h$ is

$$\delta m = \rho A \delta h \qquad (J.1)$$

This mass has a weight δmg which must be supported by the difference in gas pressure δP between the altitude h and $h + \delta h$:

$$\delta P A = -\rho A \delta h g \qquad (J.2)$$

The pressure and temperature are linked through the ideal gas law equation (G.5) which can be rewritten as

$$P = \rho k T / m \qquad (J.3)$$

where we have made use of the fact that the density is given by

$$\rho = Nm/V \qquad (J.4)$$

where m is the mass of an individual atom or molecule.

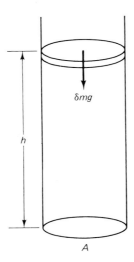

Figure A.1. Column of atmosphere

Combining equations (J.2) and (J.3) we get

$$\delta P/P = -mg\delta h/kT \tag{J.5}$$

If the atmosphere is **isothermal**, meaning its temperature does not change with altitude, equation (J.5) integrates very simply to give the equations for the exponential atmosphere:

$$P = P_0 \exp(-mgh/kT) \tag{J.6}$$

and
$$\rho = \rho_0 \exp(-mgh/kT) \tag{J.7}$$

where P_0 and ρ_0 are the pressure and density at zero altitude.

The **half-height** (h_0) of an atmosphere is the height at which the density falls by a factor two.

$$h_0 = 0.69 kT/mg \tag{J.8}$$

Appendix K: The magnitude scale

The **magnitude** scale is a modern version of an ancient system for describing the brightness of stars. It is a logarithmic scale based on the number 2.512, which happens to be $\sqrt[5]{100}$. Thus:

1 magnitude is a factor 2.512
2 magnitudes is a factor $(2.512)^2 = 6.31$
.
.
.
5 magnitudes is a factor $(2.512)^5 = 100$
10 magnitudes is a factor $(2.512)^{10} = 10\,000$
15 magnitudes is a factor $(2.512)^{15} = 1\,000\,000$

Appendix L: Boltzmann's equation

Boltzmann's equation states that when most of the transitions between energy states in atoms are the result of collisions, the relative number of molecules (n) in the lower (l) and upper (u) energy levels is given by

$$n_u/n_l = (g_u/g_l)\exp(-E/kT) \qquad (L.1)$$

where T is the temperature of the gas, E is the energy difference between the two energy levels, k is **Boltzmann's constant**, and g_u and g_l are numbers that depend on the type of atom. This relation holds for any pair of energy levels. Boltzmann's equation shows that the bigger the energy difference (E) between the lower and upper levels the smaller the fraction of atoms in the upper level will be. However, the fraction in the upper level *increases* when the temperature T is increased.

Appendix M: The Jeans criterion

The problem originally examined by James Jeans was the stability of a uniform gas under the opposing influences of gravity and thermal pressure. He found that disturbances which were larger than a critical size grew in strength, but that disturbances below the critical size were suppressed. From this theory came the idea of a **Jeans mass** as the minimum size of a cloud that could collapse.

An approximate value for the Jeans mass can be obtained by comparing the gravitational energy (E_G) and thermal energies (U) of a uniform density spherical cloud. In this simple picture the cloud will collapse if the gravitational energy exceeds the thermal energy.

Consider a cloud of radius r, and particle density n cm^{-3}. Each particle has a mass m. The volume (V) of the cloud is given by:

$$V = 4\pi r^3/3 \tag{M.1}$$

and its mass M by

$$M = nmV \tag{M.2}$$

The gravitational potential energy (E_G) of this sphere can be derived from Newton's law of gravitation and a relatively easy calculation to be:

$$E_G = 3GM^2/5r \tag{M.3}$$

where G is the gravitational constant.

The thermal energy of a monatomic gas (U) was given in equation (G.1).

$$U = 3nVkT/2 \tag{M.4}$$

where we have expressed N, the total number of particles in the cloud, in terms of the density and volume of the cloud.

If we apply the condition $E_G > U$, substitute the values of E_G and U from the above equations, and assume that the gas is a mixture of hydrogen molecules and helium atoms we can derive a critical Jeans mass M_J where

$$M_J > 6\sqrt{(T^3/n)} \; M_\odot \tag{M.5}$$

Clouds with masses above the Jeans mass have more gravitational energy than thermal energy, so they collapse. Below this mass the thermal energy is enough to keep the cloud stable.

To obtain the minimum density necessary for a cloud to form a Sun-like star out of a molecular cloud we insert $T = 10$ K and $M_J = 1$ M$_\odot$ into equation (M.5), giving the result $n = 40\,000$ cm^{-3}.

Suggestions for further reading

Two other books that introduce the interstellar medium to the general reader are *Searching Between the Stars* by Lyman Spitzer (1982, Yale University Press), and *Interstellar Matters* by Gerrit Verschuur (1989, Springer-Verlag). The first book is particularly strong in its discussion of ultraviolet astronomy and of the diffuse interstellar medium. The second book contains a history of the discovery of interstellar matter, and a series of essays on topics mainly related to radio studies of the interstellar medium.

For the reader with a strong physics background the classic graduate-student level textbooks are *Physical Processes in the Interstellar Medium* by Lyman Spitzer (1978, Wiley-Interscience) and *Astrophysics of Gaseous Nebulae* by Donald Osterbrock (1974, W. H. Freeman). Spitzer's textbook comprehensively describes the interactions of gas, dust and photons in interstellar space; the emphasis is on theory rather than observation. Osterbrock concentrates on one phase of the interstellar medium – ionized gas – but treats both theory and observation. The book has recently been revised, expanded and renamed as *Astrophysics of Gaseous Nebulae and Active Galactic Nuclei* (1989, University Science Books).

Professional astronomers usually publish their new findings in *Astrophysical Journal, Astronomical Journal, Astronomy and Astrophysics*, or *Monthly Notices of the Royal Astronomical Society*. Articles in these journals are aimed at specialists. Readers who want to explore the current frontiers of interstellar research might do better to browse through the published proceedings of two recent scientific conferences. *Interstellar Processes* (edited by D. J. Hollenbach and H. A. Thronson, 1987, Reidel) contains a splendid collection of review articles presented at a conference in Wyoming in 1986. *The Evolution of the Interstellar Medium* (edited by L. Blitz, 1990, Astronomical Society of the Pacific) is another fine collection of review articles by leading researchers; the book is based on a conference in Berkeley, California in 1989. Major review articles on particular aspects of current interstellar research appear regularly in *Annual Review of Astronomy and Astrophysics*.

Readers who appreciate the beauty as well as the physics of nebulae should turn to *Exploring the Southern Sky* by S. Laustsen, C. Madsen and R. M. West (1987, Springer-Verlag) or to *Colours of the Stars* by David Malin and Paul Murdin (1984, Cambridge University Press). Both books combine magnificent color photography with scientific texts written by professional astronomers.

The monthly magazine *Sky and Telescope* frequently contains articles on interstellar matter written by professionals but aimed at amateur astronomers. The illustrations in *Sky and Telescope* are often superior to those published in the technical press.

Picture acknowledgements

Figures 1.1, 1.3, 1.4: © Anglo-Australian Observatory
Figure 1.5 (upper and middle): courtesy of the COBE Science Working Group and NASA
Figure 1.5 (lower): © Lund Observatory
Figure 2.4: courtesy MMT Observatory
Figure 3.1: California Institute of Technology
Figure 3.4 is adapted from work by M. Heydari-Malayeri, E. van Drom, and P. Leisy in *Astronomy and Astrophysics* **240,** 481 (1990)
Figure 3.5: Lick Observatory Photograph
Figure 4.2: courtesy Remo Tilanus
Figure 4.3: Max-Planck-Institut für Radioastronomie, Bonn, Germany
Figure 4.4: courtesy NRAO/AUI
Figure 4.5: courtesy Christine Jones and William Forman
Figure 4.7: courtesy Gart Westerhout
Figure 4.8: adapted from work by R. D. Davies and E. R. Cummings in *Monthly Notices of the Royal Astronomical Society* **170,** 95 (1975)
Figure 4.9: courtesy Carl Heiles
Figure 5.2: courtesy of Canada-France-Hawaii Telescope corporation: photographer: Tom Gregory
Figure 5.3: © Anglo-Australian Observatory
Figure 5.4: courtesy of Canada-France-Hawaii Telescope Corporation
Figure 5.5: © European Southern Observatory
Figure 5.7 was prepared by Jeffrey Goldader from data supplied by Jack Welch and John Dreher
Figure 5.8: adapted with permission from a figure by Y. M. and Y. P. Georgelin in *Astronomy and Astrophysics* **49,** 57 (1976)
Figure 5.9: adapted from a figure kindly supplied by Ronald J. Reynolds
Figure 6.1: Lick Observatory Photograph
Figure 6.2: adapted from work by Donald Morton published in *Astrophysical Journal* **197,** 85 (1976)
Figure 7.2: courtesy Jeffrey Goldader and the author
Figure 7.3: © Royal Observatory, Edinburgh and Anglo-Australian Telescope Board
Figure 7.5: Courtesy Klaus Hodapp: originally published in *Astronomy and Astrophysics* **172,** 304 (1987)
Figure 7.6: produced by the Infrared Processing and Analysis Center, which is funded by NASA under contract to the California Institute of Technology and the Jet Propulsion Laboratory

Figure 7.8: courtesy NASA
Figure 7.9: adapted from work by R. W. Russell, B. T. Soifer and S. P. Willner published in *Astrophysical Journal* **220,** 568 (1978)
Figure 8.2: adapted with permission from work by Lyman Spitzer and Edward Jenkins published in *Annual Review of Astronomy and Astrophysics* **13,** 133 (1975)
Figure 8.3: published by Ron Garden, Adrian Russell, and Michael Burton in *Astrophysical Journal* **354,** 232 (1990)
Figure 8.4: figure supplied by Geoff Blake, who also compiled the data in Table 8.1
Figure 8.5: courtesy Ron Snell and Peter Schloerb, published in *Astrophysical Journal* **283,** 129 (1984)
Figure 8.6: © Harvard College Observatory
Figure 8.7: courtesy R. J. Maddalena, M. Morris, J. Moscowitz, and P. Thaddeus, adapted from an article in *Astrophysical Journal* **303,** 375 (1986)
Figure 8.8: produced by the Infrared Processing and Analysis Center, which is funded by NASA under contract to the California Institute of Technology and the Jet Propulsion Laboratory
Figure 8.9: supplied by Dennis Downes of IRAM
Figure 8.10: supplied by Thomas Dame
Figure 8.11: adapted from a figure produced by Nick Scoville and David Sanders
Figure 8.12: taken from a paper by a group led by Reinhard Genzel published in *Astrophysical Journal* **247,** 1039 (1981)
Figure 9.3: courtesy European Space Agency
Figure 9.4: courtesy NASA
Figure 9.5: courtesy C. G. T. Haslam
Figure 9.8: courtesy D. S. Mathewson, based on a figure in *Memoirs of the Royal Astronomical Society* **74,** 139 (1970)
Figure 9.9: courtesy Rainer Beck
Figure 10.1: courtesy of Canada-France-Hawaii Telescope Corporation
Figure 10.2: taken from a paper by E. P. Ney and others in *Astrophysical Journal* **198,** L129 (1975)
Figure 10.3: courtesy Canada-France-Hawaii Telescope Corporation and Laird Thompson
Figure 10.4: courtesy Fred Seward
Figure 10.5: courtesy NRAO/AUI
Figure 10.6: © European Southern Observatory
Figure 11.2: Lick Observatory Photograph
Figure 12.1: courtesy George Herbig from a paper published in *Astrophysical Journal* **214,** 747 (1977)
Figure 12.2: from a paper by B. A. Wilking, C. J. Lada and E. T. Young, in *Astrophysical Journal* **340,** 823 (1989)
Figure 12.5: from work by F. C. Adams, C. J. Lada and F. H. Shu, published in *Astrophysical Journal* **312,** 788 (1987)
Figure 12.7: courtesy B. A. Smith and R. J. Terrile
Figure 12.8: courtesy P. F. Goldsmith, R. L. Snell, M. Hemeon-Heyer and W. D. Langer from a paper published in *Astrophysical Journal* **286,** 599 (1984)
Figure 12.9: courtesy F. H. Shu, F. C. Adams and S. Lizano, published in *Annual Review of Astronomy and Astrophysics*, **25,** 23 (1987)
Figure 13.1: courtesy P. B. Hutchinson
Figure 13.2: reproduced, with permission, from the *Annual Review of Earth and Planetary Sciences vol 13* © 1985 by Annual Reviews Inc
Figure 13.3: Lick Observatory Photograph
Figure 13.4: High Altitude Observatory and Southwestern at Memphis

Figure 13.6: courtesy NASA
Figure 15.1: reproduced from *Monthly Notices of the Royal Astronomical Society* **178,** 577 (1977)
Figure 15.2: courtesy Christine Jones and William Forman
Figure 15.3: courtesy Frazer Owen
Figure 15.4: adapted from a figure by P. J. Young, W. L. W. Sargent, A. Boksenberg, R. F. Carswell and J. A. J. Whelan published in *Astrophysical Journal* **229,** 891 (1979)

Index

absolute zero 28
absorption
  of electromagnetic radiation 13
  self 91
absorption lines
  from interstellar gas 59
  from quasars 174
  in stellar outflows 114
  saturation of 59
  solar 23, 62
  spectrum 23
  21-cm 39
abundances of elements 55–63
  cosmic 62
  in cosmic rays 101, 105
  helium 55
  in H^+ regions 56
  in intergalactic medium 170, 175
  in interstellar medium 60
  in Solar System 61
Alfvén waves 141
ambipolar diffusion 141
argon 26, 82, 163
associations, stellar 136
asteroids 2, 156
astronomical unit (a.u.) 180
atmosphere
  of Earth 5, 162–8
  and gravity 190
  windows in 14, 72
atomic hydrogen
  in galaxies 31
  in interstellar medium 30–42, 125–34
  in Milky Way Galaxy 35–8, 40, 95
atomic number 19
atomic structure 19
atomic weight 19
Aurora Borealis 157

Balmer series 48
baryons 177
beryllium 61, 101
Big Bang 111, 176
bimodal star formation 148
bipolar flows 83, 147
bipolar nebulae 116
black body 28, 69
black-body radiation 28, 188
black holes 177
blueshift 24
Boltzmann's constant 187, 192
Boltzmann's equation 82, 192
boron 61, 101
bremsstrahlung 102
bubbles 129, 132

calcium 59
carbon
  abundance 61
  in grains 75
  infrared emission line 127
  origin of 112
carbon monoxide (CO)
  abundance of 85
  bipolar flows 148
  circumstellar 114
  and dust grains 76, 123
  isotopes 91
  in Milky Way Galaxy 94
  in molecular clouds 88
  transitions of 80
chemical composition 55
chemistry 84
circumstellar shells 71, 114, 125
cirrus 70
clouds
  classification of 125
  collapse of 135, 139, 194
  cool 40, 83, 126
  evidence for 39
  destruction of 131
  intergalactic 175
  origin of 127, 130
  rotation of 145
  shapes of 41, 131–2
  stability 139
  theoretical 130
  transparent 81
  see also nebulae, molecular clouds
clusters
  of galaxies 2, 170–4, 176
  of stars 136, 139, 176
COBE (Cosmic Background Explorer) 7, 73
collisional de-excitation 51, 56
collisional excitation 33, 56, 82, 129
collisional ionization 43, 54, 128
computers in astronomy 144
comets 2, 155, 160
continuum radiation 22, 28
cooling flow 173
Copernicus satellite 58, 81
corona of Sun 157
coronal gas 54, 128–34
  in supernova remnants 120
COS-B satellite 101, 104
cosmic rays 11, 98–105
  composition of 100–1
  detectors 103
  energies 100
  in the Galaxy 99
  heating by 127
  ionization by 87
  in molecular clouds 87
  motions of 99
  origin of 105
  in upper atmosphere 167
cosmology 176
Crab Nebula 101, 118
critical density 176

dark clouds: see nebulae, molecular clouds
dark matter 8, 34, 175
density 5
depletion 62, 77, 123
deuterium 82, 111, 177

differential rotation 32
diffuse interstellar bands 76
disks
   of galaxies 2, 31, 36
   around stars 146
dissociation 26, 81
   in Earth's atmosphere 164
distances 3
Doppler effect 24, 186
   galaxy rotation 31, 36, 52, 94
   nebula expansion 89, 116
   outflows 114, 146
   widths of spectral lines 41, 59, 141
dust clouds: *see* nebulae, molecular clouds
dust grains 64–78
   catalysts 87
   composition of 75
   destruction of 124
   formation of 123
   in H^+ regions 72
   infrared emission from 69
   intergalactic 175
   interplanetary 152–6
   in molecular clouds 26
   photoelectric effect 127
   in planetary nebulae 115
   reddening by 64
   shapes 74
   sizes 74, 152
   temperature 70, 77
   see also PAH molecules
dust shells 71, 114
dust-to-gas ratio 78

Einstein X-ray satellite 121, 172
electromagnetic spectrum 10, 15
electromagnetic waves 10
elements 19
   abundances 55–63
   heavy 19, 56, 113
   light 19
   origin of 111
   table of 185
emission lines 22
   coronal gas 128
   H^+ regions 48
   infrared 52
   from planetary nebulae 115
   radio 52
   spectrum 23–4
   stellar outflows 114
   from supernova remnants 120
   from T Tauri stars 137
   21-cm 30
endothermic reactions 87, 113
energy levels
   hydrogen 21, 31, 49
   oxygen 49
excited states 20, 84
exothermic reactions 113

extinction 5, 64
   curve 65

Faraday rotation 108
fluorescence 84
flux freezing 105
forbidden lines 33
   in H^+ regions 50
   in planetary nebulae 115
formation of stars 135–51
free–free emission 51, 122
frequency 12, 184

galactic center 36, 95, 170
galactic fountain 134, 170, 173
galactic halo 133
galactic plane 36, 104
   thickness of 38, 105, 133
galaxies 2
   atomic hydrogen in 31
   clusters of 2, 170–4
   masses of 32–4
   radio 172–3
   rotation of 189
Galaxy, the 2
   see also Milky Way Galaxy
gamma rays
   from Milky Way Galaxy 101
   origin of 102
   telescopes 103
   wavelengths of 11
gas
   atomic 30–42
   composition 55–63
   cooling 126
   coronal 54, 128–34
   density of 5, 40, 53, 81, 127
   heating 126
   intergalactic 169–75
   interplanetary 156–9
   ionized 43–54
   molecular 79–97
   motions 41
   phases of 30, 40, 81, 127
   pressure 27
   temperature of 40, 50, 53, 82, 89, 126
gas-to-dust ratio 78
geospace 2, 162–8
graphite 65, 75, 77, 124
gravity
   and atmospheres 162, 190
   and cloud collapse 139, 194
   and dark matter 175
   law of 138
   in stars 114
ground state 20

H^0 regions 30
   see also gas, atomic
H_2 molecules
   formation of 88

   infrared transitions 83
   ultraviolet transitions 81
H_2 regions 30
   see also molecular clouds
H^+ regions 30, 43–54
   abundances in 57
   distances to 52
   galactic distribution 53, 95
   heating of dust in 72
   infrared emission from 52, 72
   light from 48
   obscuration of 66
   radio emission from 51, 105
   and star formation 149
   temperature of 50
half-height 38, 172, 191
halo 133
$H\alpha$ line 49, 54
heliopause 159
helium 56
   abundance 55
   atomic structure 20
   in Earth's atmosphere 164
   origin of 111
   recombination lines 23, 49, 52, 55
   wake 159
Herbig–Haro objects 147
HI regions: *see* gas, atomic
high-velocity clouds 134, 151, 170
HII regions: *see* H^+ regions
Hubble Space Telescope (HST) 17, 58
hydrogen
   atomic structure 19, 31
   in Earth's atmosphere 164
   recombination lines 23, 48, 52
   see also H^+ regions, molecular clouds, atomic hydrogen
hydroxyl (OH) 95, 110
   circumstellar 114

ice 65, 76, 124, 143
infrared radiation
   detection of 72
   from dust 69, 114
   from Milky Way 7
   from protostar 143
   wavelengths of 10
intergalactic medium 2, 169–78
   diffuse 175
International Ultraviolet Explorer (IUE) 58
interplanetary medium 2, 152–61
interplanetary scintillation 158
interstellar clouds: *see* clouds
interstellar dust: *see* dust grains
interstellar gas: *see* gas
interstellar medium 2, 126
   diffuse 5, 126
   discovery of 3
   fate of 150

local 132
mass of 7, 33, 38, 130
origin of 111–24
ion–molecule reactions 87
ionization 19, 43
collisional 43, 54, 128
in Earth's atmosphere 164
front 45
potential 21, 49, 56, 60, 185
ionized gas 43–54
see also H^+ regions
ionizing photons 44
ionosphere 164
ions 19
molecular 84
IRAS Infrared Astronomy Satellite 70, 73, 153
isotope 19, 91

Jeans criterion 139, 194

kinetic energy 27, 100, 129
Kuiper Airborne Observatory 73

laser 95
light
speed of 11, 99
wavelength of 11
light elements 19
lithium 61, 101, 111, 137
Local Group 2, 172
local interstellar medium 132, 158
Lyman-α 49, 81
backscatter 158
quasar absorption lines 174
saturation 59

magnetic braking 145
magnetic fields 99, 140
in galaxies 106, 109
measuring 106
and polarization 75, 108
magnetic forces 99, 140
magnetic pressure 140
magnetism
ambipolar diffusion 141
flux freezing 105, 140, 145
interplanetary 159
interstellar 75, 105–10
synchrotron radiation 104
magneto-hydrodynamic waves 141
magnetopause 168
magnetosphere 166
magnitudes 64, 192
masers 95–97, 146
mass loss 113
meteorites 61, 153
meteors 153
metric system 180

microwaves 10
Milky Way Galaxy 2
age 12
atomic hydrogen in 35, 38
differential rotation 32, 106
extinction across 64
gamma rays from 101
H^+ regions in 53
halo 133
Hα emission 54
infrared emission from 7
light from 7, 48
molecular clouds in 94
radio emission from 104
rotation of 2, 12, 37, 145, 189
millimeter waves 10, 92
missing mass 176
models
computer 142
galactic 37
protostars 142
stellar 112
of Universe 3
molecular clouds 5, 26, 79–97
collisions 148
cores 137
dense 88, 126
density in 89
in the Galaxy 94
gamma ray from 102
giant 5, 79, 89
life inside 93
masses 91
rotation 145
sizes 89
spectrum 86
temperatures 81, 89
transparent 81, 130
young stars in 137
molecular weight 163
molecules 26
formation of 87
interstellar 85
spectroscopy of 79
transitions of 26, 79

nebulae 3
bipolar 116
bright 5, 6, 45, 46; *see also* H^+ regions
dark 5, 6, 88, 137; *see also* molecular clouds
planetary 46, 47, 115
reflection 5, 18, 67, 116, 152
spectra of 23
neon 23, 52, 57
neutron stars 118
novae 122
nucleosynthesis
in Big Bang 55, 111
in stars 112
in supernovae 118

OH: *see* hydroxyl
Oort cloud 160
organic molecules 84
organic refractories 75
Orion Molecular Clouds 83, 89, 90
Orion Nebula 46, 56, 89, 90
outflows 146
oxygen
abundance of 57
in Earth's atmosphere 163
energy levels 49
forbidden lines 50
origin of 112
spectral lines 23
ozone 164

PAH molecules 27, 76, 88, 127
phases 30, 40, 81, 127
photoelectric heating 127
photography 47
photoionization 43, 128
of intergalactic medium 174
photons 12
ionizing 44
pion decay 101
Planck's constant 12
Planck's law 188
planetary nebulae 46, 47, 115
Planets 2
formation of 146
plasma 43, 172
see also H^+ regions
Pleiades 18, 67
polarization
of scattered light 68
of starlight 74, 108
of synchrotron radiation 105, 109
polycyclic aromatic hydrocarbons: *see* PAH molecules
pressure
degeneracy 117
gas 27, 126, 139, 187
magnetic 140
protostars 142, 145
pulsars 53, 109, 119

quantum theory 12, 20, 79
quasars 93, 174

radial velocity 24, 41
radiation 9, 28
see also electromagnetic waves
radiation pressure 115, 155–8
radio astronomy 34, 92, 165
radio communication 165
radio spectroscopy 26
radio wavelengths 10
recombination lines 48, 115
reddening 5, 65, 74
redshift 24
refraction 14

saturation 59
scattering
    by dust 14, 18, 66
    of electromagnetic radiation 14
scientific notation 179
self-absorption 91
shock waves 42
    and chemistry 87
    and coronal gas 129
    and cosmic rays 105
    destruction of dust 124
    H_2 lines from 83
    heating of interstellar gas 129
    heliopause 159
    in protostars 142
    and star formation 141
silicates 65, 75, 124, 143, 154, 156
silicon carbide 77, 156
solar backscatter 132, 158
Solar System 2
    abundances in 61
    dust in 152–6
    edge 159
    formation of 146
    gas in 156–61
solar wind 100, 113, 156
solids 27
    see also dust grains
sound waves 42
space travel 2
spectral features 37
spectral lines
    absorption 23
    broadening 22, 41, 59
    emission 22
spectroscopy 22, 25
spiral arms 2, 31, 37, 106, 149
star formation 72, 135–51
    bimodal 148
    rate 136
    total 150
starbursts 171
stars 1
    binary 122, 145
    birth 135–51
    black dwarf 117
    brown dwarf 177
    carbon-rich 123, 156
    clusters of 136
    dust shells 71
    evolved 112
    exciting 44, 45
    formation 135–51
    giant 71, 97
    lifetimes 136
    main-sequence 45, 112

mass loss from 113
neutron 118
OB 45, 66, 114, 136, 148, 150
outflows 148
oxygen-rich 123
pre-main-sequence 143
red giant 112, 123
T Tauri 137, 143
white dwarf 46, 113, 117
young 136–7
statistical mechanics 29
Stefan's law 188
stellar winds 105, 113, 146, 148
stimulated emission 96
Sun 1
    abundances in 61
    continuous radiation 28
    corona of 157
    helium production in 112
    luminosity of 182
    mass of 181
    rotation of 145
    spectrum 23
    sunspots 166
superbubbles 149
supernova remnants (SNRs) 119
    coronal gas 129
    and cosmic rays 105
    dust formation 125
    emission lines from 120
    gamma rays from 101
    radio emission from 105, 120
    X-ray emission from 120
supernovae 117
    and cosmic rays 105
    nearby 132
    nucleosynthesis 114
    and star formation 149
    type I 123
    type II 118
synchrotron radiation
    from Milky Way Galaxy 104
    polarization of 108
    from supernova remnants 119
    X-ray 122
Système International 180

taxonomy 125
telescopes
    gamma-ray 103
    infrared 72
    millimeter-wave 92
    radio 34, 92
    ultraviolet 57
    visible light 16
    X-ray 121

temperature 27, 69, 187
    of atomic gas 40, 126
    of Earth's atmosphere 164, 166
    equilibrium 69
    of intergalactic gas 172
    of ionized gas 50
    of molecular gas 81, 89
    scales defined 182
thermal energy 50, 187, 194
thermal motions 27, 41
thermal radiation 28, 73, 188
tidal bridges 170
transitions
    collisional 21, 82
    downward 21
    electronic 21, 79
    forbidden 33, 50
    of molecules 79
    permitted 33
    rotational 80
    spontaneous 21, 32
    stimulated 21, 39
    upward 21, 57
    vibrational 79
tunnels 129
21-cm line 30, 110, 132, 169

ultraviolet astronomy 57
ultraviolet fluorescence 84
ultraviolet radiation 11
Universe
    fate of 176
    origin of 111
    study of 169

van Allen belts 167
virial theorem 175
Voyager spacecraft 3, 160

warm ionized medium 52, 126, 130
warm neutral medium 40, 126, 130, 132, 158
water vapor masers 96
wavelength 10, 184
Wien's law 28, 188

X-rays
    from coronal gas 128
    from galaxy clusters 172
    heating 127
    from supernova remnants 120
    wavelength of 11

Zeeman effect 108
zodiacal light 152